The Bible Code Princess Diana and Dodi Fayed

Diana and Dodi, their final moments alive together.

This book was achieved using the results of research utilizing a Bible Code computer program. The Bible Code is a scientific mathematical theory, and like all theories, old or new, may be proven to be true or false as the world evolves to an ever higher technological and spiritual plane.

Written by
Pamela Lillian Valemont
BA Film & TV Production

ISBN 978-1-105-99138-7
 This book has a companion book: _The Bible Code and Australia The Garden of Eden_, also available at http://www.lulu.com/spotlight/pamelavalemont

Eternal Flame at the Pont d'Alma – scene of an "assassident?"

Synopsis

What happened to Diana and Dodi in the Alma Tunnel on the night of August 31, 1997?

Was Diana six weeks pregnant? Was she about to get engaged and married to Dodi? Had Dodi purchased an engagement ring for her? Was he planning to present it to her that night?

Was Andanson an SIS operative, trained to use his white Fiat Uno to tip the Mercedes car, causing it to career into the 13th pillar in the tunnel? Why was he found, deep in a forest, beheaded in his burnt out, keyless, locked car, with a bullet wound in his head?

Is Tomlinson, ex-SIS agent, right when he says the Mercedes "car accident" was a classically planned SIS method of execution, one he knew was being planned to assassinate a former world Serbian leader?

What was the bright light witnesses saw in the tunnel that night? Was it a strobe light used to blind the driver? What was the loud bang they heard *before* the crash?

Did Diana tell a medic that she wished to be buried with Dodi as she lay dying? Was Diana a half-sister to Jemima Khan? Had Diana been planning to convert to the Muslim faith to marry as Jemima did when she married Imran Khan?

The Bible Code remained elusive to Sir Isaac Newton and Albert Einstein, though they tried for years to uncover it. Now unlocked and opened to us since the timely invention of the computer, all these questions, astoundingly, may be answered in a book 3,000 years old, hidden in encoded equidistant letter matrices.

Section 1. Why I wrote the book, and how my life as a poet and numerologist played a big part.

Recently I became intrigued with the findings of Eliyahu Rips, the world famous Israeli mathematician, and Michael Drosnin, USA political and investigative journalist, outlined in The Bible Code, written by Drosnin. Although some sources say that Rips has distanced himself from the findings of Drosnin, nonetheless Rips is credited with being the discoverer of The Bible Code. All Rips' research has been in Hebrew, using the Old Testament, in the Torah. Their findings were corroborated by a former USA government intelligence code breaker. In the late seventies, Rips began looking, with the help of a computer, for codes in the Torah.

(References: (a) In 1994, Rips, together with Doron Witztum and Yoav Rosenberg, published a seminal article in the journal Statistical Science, "Equidistant Letter Sequences in the Book of Genesis", which claimed the discovery of encoded messages in the Hebrew text of Genesis. This, in turn, was the inspiration for the 1997 book The Bible Code by journalist Michael Drosnin. While Rips originally claimed that he agreed with Drosnin's findings, he later distanced himself from his interpretations. Since Drosnin's book, Bible codes have been a subject of controversy, with the claims being criticized by Brendan McKay the Australian mathematician, and others. http://en.wikipedia.org/wiki/Eliyahu_Rips

(b) Drosnin, Michael (1997). The Bible Code. Simon & Schuster. Drosnin, Michael (2002). The Bible Code II: The Countdown. Viking Books. Drosnin, Michael (2006). The Bible Code III: The Quest. Weidenfeld & Nicolson.)

Because I am an English speaking poet and numerologist, I decided to use the text in my native tongue. One of the criticisms levelled at Drosnin's and Rips' findings, is that they refer to ordinary, that is linear Biblical text that crosses or intersects with the matrix, as being of especial value, helping to strengthen and support the code findings in their eyes. Many mathematicians disagree: they claim that linear text is of no use at all in establishing credibility, that only the matrix itself could be used as evidence of the veracity of the code. This is something you will have to decide for yourselves. In conducting this new research, I merely present my findings to you.

My modus operandi has always been that I use my (psychic) intuition, which I follow up with numerological analysis. This method has been useful in the past. Here is a cut and paste of an email I sent to Natasha Dickins, of The Women's Weekly, on November 19, 2003.

I sent it to her at Australian Consolidated Press. I predicted the capture of Saddam Hussein to within 16 days of accuracy. He was captured on 13th December, 2003. My estimate was "before the 28th of November, 2003". Remember of course, that my prediction was made at a time when no one thought he would be captured then.

Within twenty minutes of the announcement of the Princess of Wales, Diana's death, I sat down to write an article. In it, I explained there was a possibility that her death could have been deliberately engineered, because of the fact that she was in a 34/7 year.

Her last birthday before the accident fell on 1st July, 1997 = 34/7 Personal Year. (1+7+(1+9+9+7=26))= 34/7 In my experience, some young people attempt suicide during this especially difficult year for them emotionally, that comes hard upon the Master 33/6 year they have just left behind them.

During the 33/6 year, they are often separated from those they love. Permanent partings and new homes can result. This can cause great suffering. The 34/7 year that follows is sometimes associated with disputes and estrangements from family, which can, in the extreme, lead to physical violence. Divorce settlements are common also during this time.

S U I C I D E
1 3 9 3 9 4 5 =34/7

M U R D E R
4 3 9 4 5 9 = 34/7

Although for over thirty years I have written numerology columns for Australian newspapers, pre-recorded hour long programs, done endless talk back radio programs and television appearances, all persona non gratis, I have never yet had one journal, radio or television station offer to pay me for even one of my articles or appearances.

Not long after I sent my article to AAP, suspicions arose as to the circumstances surrounding the deaths of Diana and Dodi. They have continued to this day. Yet New Idea have had no problems publishing interviews with Richard Tomlinson, ex MI6 agent, and I assume paying him for his story, even though his contentions were far more extreme than were mine.

----- Original Message -----
From: Pamela Valemont
To: ndickins@acp.com.au
Sent: Wednesday, November 19, 2003 3:13 PM
Subject: Re: Numerology of killing of John Lennon

(Reference: Numerology in Action in Our World.
(http://www.lulu.com/spotlight/pamelavalemont and ebook sites.)

Dear Natasha,

In regard to our conversation just now, here is the said journalistic article. Also, I am resending the Princess Diana/Prince William one as well.

I will also include the one I sent you in August for Pauline Hanson, showing when I believed her appeal would take place, between 27/10/03-27/11/03) which it did, the possibility of a hurried decision in her favour, overturning the usual slow process of the law, and the danger of her becoming depressed and contemplating suicide - all of which eventuated.

I feel now, that, given the Master 38/11 year she will be going into at her next birthday in on 27th May, 04, she will receive a massive compensation payout. I also feel that, based on the birthday of 28th April for Saddam Hussein, that he will be taken prisoner before his next birthday in 2004.

At the moment, I am extremely worried about the fact that he is still in a Master 47/11 Personal Month from 28/10/03-28/11/03, which presides over combustible things, including bombs, and since he has, in the past, been responsible for the burning of villagers with white phosphorous, I am concerned that he might try something like that again.

This would be an excellent month for the Americans to take him prisoner, as he is also passing through a Justice month, when the universal law could catch up with him. There is of course, the distinct possibility that he will be the target of a personal attack, either an attempt to kill him, or take him prisoner, before the 28th of November has passed.

Kind Regards,
Pamela Valemont BA Numerologist. (End of Quote)

Below is my interpretation of the number 34/7 and the things it rules.

Section 2. How being an Australian played a part in my decision to conduct this biblical research.

Because Australia is a member of the British Commonwealth, I think it is fair to say that perhaps the concern generated by the controversy surrounding the deaths of Princess Diana and Dodi Fayed touched us more than many other countries around the world.

Of particular concern to Australians have been the charges levelled at several members of the royal family, implicating them in the deaths of Diana and Dodi. Those targeted for these charges have included Queen Elizabeth, Prince Philip, Prince Charles, heir to the throne, and Prince Michael of Kent.

The ongoing court cases conducted by Mohamed Al Fayed in the wake of the death of his son and Diana raise serious concern for many Australians. They are worried on two counts. First, if the charges were proven to be true, then

Australians as members of the British Commonwealth are subjects to a corrupt royalty.

Second, if the charges are false, and so far they have proven to be in the courts, then we as Australians have a right to feel outraged. We also naturally feel great compassion for the royal family because of this dreadful and humiliating saga they have been forced to undergo. It has been a lengthy, draining process which must grieve and cost them dearly in terms of their emotional wellbeing, one that would in turn naturally impinge on their overall mental and physical wellbeing.

As I write this ebook, Queen Elizabeth II is celebrating 75 years on the English throne – her Diamond Jubilee. Prince Philip has been unable to attend the celebrations due to a urinary tract infection. He will be 90 on the 10th of June, and such illnesses are not uncommon at his age, but they are exacerbated by stress. He has had much more than his fair share of stress recently, and with the outcome of the second French court inquiry into Diana's and Dodi's deaths currently adjourned, this elderly gentleman must be undergoing a great deal of emotional anguish, not only on his own behalf, but out of concern for his family members.

Mosaic of Queen Elizabeth, by Helen Marshall, creating a timeline from her ascension to the throne on 2nd June, 1953 to her Diamond Jubilee in 2012, displayed at the Towner Gallery, Eastbourne, London. Elizabeth as a child below, and the Royal Crown.

Motive

In **1998** the UK abolished the Death Penalty entirely, but in **1965** the **Death Penalty was reserved for "Treason, piracy, and committing adultery with the wife of the monarch's eldest son.**" At that stage, Prince Charles was only 17 years old, having been born in November, 1948. Guess they were getting in early. If anyone doubts Princess Diana's death was anything other than accidental, you only have to look at the fact that she was killed in **1997**. The following year, **1998**, the Death Penalty was abolished entirely in the UK, but in the year Diana was killed, this law was still in force. Although Diana had divorced Prince Charles, we can easily see that this law was passed to protect the lineage of the Royal Family. So, any person who committed a sexual act with Diana, that is, the wife of the eldest child of the monarch, (liable of course to result in pregnancy) would be put to death by the law of the land.

The rumour that Diana was pregnant would definitely give them more than enough reason to put Dodi to death. Perhaps getting rid of Diana was a necessity for them, as she was young, and still of child bearing age at the time of her divorce from Charles, the future monarch. It also gives weight to the charge made by Captain Hewitt (who was rumoured to be the father of Harry) that he was threatened with death if he did not give Diana up. I tend to believe he definitely was not, or he would have been killed years ago. How sexist was this law?

Down through the ages, kings have been known to sire illegitimate children, yet if the wife of a man in line to the throne even engaged in sexual intercourse with a man other than her husband, her lover would be executed. When you look at things this way, Diana really was testing and defying the existing law in coming out and revealing the identities of the lovers in her life before she died.

She was in fact, exposing them to the Death Penalty, should the UK have gone ahead and implemented that (then) existing law. Perhaps they were unwilling to prosecute Christian white men, but you can well believe they would have no hesitation at all in executing an Egyptian-born Muslim man, particularly if he had sired a child who was half-brother or sister to the English monarch of the future, Prince William. Diana was much loved by the people. Rather than take the matter to court, where they might have met with the weight of public opposition, they decided to simply assassinate them both on the sly. One of the last pictures of Diana alive below, taken with her sons.

Princess Diana with sons Harry and William and a family friend on Barbuda in 1997.

Section 3.The controversy surrounding the deaths of Diana, Princess of Wales and Dodi Fayed on the roadway under the Pont de l'Alma, known as the Alma Tunnel.

The Pont de l'Alma where Diana died and Althorp, where she grew up. Pont de l'Alma (Alma Bridge in English) is an arched bridge in Paris, crossing the Seine, named to commemorate the Battle of Alma during the Crimean War, in which the Franco-British alliance achieved victory over the Russian army on 20th September, 1854.

The initial construction of the bridge took place between 1854 and 1856. It was inaugurated by Napoleon III on 2nd April 1856. At the time, each of the four piers of the bridge was decorated with a military statue: a sculpture of a zouave and a grenadier, by Georges Diébolt; a skirmisher and an artilleryman, sculpted by Arnaud.

For Parisians, the statue of the Zouave soldier serves as a floodwater level indicator. Access to the footpaths by the river embankments is closed when the Seine's level reaches the feet of the Zouave, and when the water hits the statue's thighs, the river is considered unnavigable. During the great flood of the Seine in 1910, above, the level eventually reached to the shoulders of the Zouave - the poor fellow must have looked prophetically like he was going down in the Titanic - which great maritime disaster happened just two years later.

Dodi's father erected two memorials to Dodi and Diana at Harrods, which famous store he was forced to sell, in order to pay the massive legal costs of the court case he mounted trying to prove Diana's father-in- law, the Duke of Edinburgh, Philip Mountbatten, was behind a plot to kill Diana.

He later reduced the charges to the profoundly lesser claim that Philip had wanted Diana out of the way, that this was known to the SIS/MI6, who had then taken matters into their own hands and disposed of her. In other words, that Philip was not directly responsible for the accident, or consequently, the deaths of the pair.

Above: Memorials to Diana and Dodi, erected at Harrods by Dodi's father. Below: Dodi, an Englishman of Egyptian descent, of the Muslim faith, who wanted to marry an English Princess of the Anglican faith. Were they just too naïve to see the danger involved?

Matches for: Diana Fayed die heart tear trauma

Match: 1 of 1 Skip to Match

ELS (Equidistant Letter Spacing) Matrix

```
DSINTHEMIDSTOFHIMTHEYSHALLSANCTIFYMYNAMEYEATHE
YSHALLSANCTIFYTHEHOLYONEOFJACOBANDSHALLSTANDIN
AWEOFTHEGODOFISRAELTHEYALSOTHATERRINSPIRITSHAL
LCOMETOUNDERSTANDINGANDTHEYTHATMURMURSHALLRECE
IVEINSTRUCTIONWOETOTHEREBELLIOUSCHILDRENSAITHJ
EHOVAHTHATTAKECOUNSELBUTNOTOFMEANDTHATMAKEALEA
GUEBUTNOTOFMYSPIRITTHATTHEYMAYADDSINTOSINTHATS
ETOUTTOGODOWNINTOEGYPTANDHAVENOTASKEDATMYMOUTH
TOSTRENGTHENTHEMSELVESINTHESTRENGTHOFPHARAOHAN
DTOTAKEREFUGEINTHESHADOWOFEGYPT!THEREFORESHALL
THESTRENGTHOFPHARAOHBEYOURSHAMEANDTHEREFUGEINT
HESHADOWOFEGYPTYOURCONFUSIONFORTHEIRPRINCESARE
ATZOANANDTHEIRAMBASSADORSARECOMETOHANESTHEYSHA
LLALLBEASHAMEDBECAUSEOFAPEOPLETHATCANNOTPROFIT
THEMTHATARENOTAHELPNORPROFITBUTASHAMEANDALSOAR
EPROACHTHEBURDENOFTHEBEASTSOFTHESOUTHTHROUGHTH
ELANDOFTROUBLEANDANGUISHFROMWHENCECOMETHELIONE
SSANDTHELIONTHEVIPERANDFIERYFLYINGSERPENTTHEYC
ARRYTHEIRRICHESUPONTHESHOULDERSOFYOUNGASSESAND
THEIRTREASURESUPONTHEHUMPSOFCAMELSTOAPEOPLETHA
TSHALLNOTPROFITTHEMFOREGYPTHELPETHINVAINANDTON
OPURPOSETHEREFOREHAVEICALLEDHERRAHABTHATSITTET
HSTILLNOWGOWRITEITBEFORETHEMONATABLETANDINSCRI
BEITINABOOKTHATITMAYBEFORTHETIMETOCOMEFOREVERA
NDEVERFORITISAREBELLIOUSPEOPLELYINGCHILDRENCHI
LDRENTHATWILLNOTHEARTHELAWOFJEHOVAHTHATSAYTOTH
ESEERSSEENOTANDTOTHEPROPHETSPROPHESYNOTUNTOUSR
IGHTTHINGSSPEAKUNTOUSSMOOTHTHINGSPROPHESYDECEI
```

Verses in Matrix

The Book of the Prophet Isaiah 29:23-30:10

Among the injuries Dodi suffered were a rupture in the isthmus of the aorta and a fractured spine. The autopsies showed that Diana's heart had been displaced from the left to the right side of the chest, which tore the pulmonary vein and the pericardium. Their injuries in common related to the heart. The above matrix makes mention of this, without reliance on the ordinary text.

The one below however, has "heart" appearing in the ordinary text, intersecting with "Fayed" and "seat" of the matrix. Neither he nor Diana were wearing seat belts. Why they were not, when Diana in particular was said to be very concerned always to do so, has been a source of much speculation. Was Diana's seat belt tampered with? It was found to be broken after the crash, but analysts dismissed this, as having been caused by the force of the accident, which is of course, quite possible.

Harrods

Mohamed on his luxury yacht with youthful family members. Diana and Dodi on their last holiday together.

When conducting searches (in the system I used) that included the names of those fatally injured in the crash, neither the name Dodi, nor his birth name of Emad came up, but the surname Fayed came up repeatedly in connection with Diana. The name of Dodi comes up in other Bible Code Matrix systems, but the reliability of these systems in other respects is questionable.

Matches for: Fayed rear car seat Alma tear spine heart kill

Match: 1 **of 1** Skip to Match

ELS (Equidistant Letter Spacing) Matrix

```
ATIONTOTHECHILDRENOFMINEOWNMOTHEREVENYOUNGCHIL
DRENDESPISEMEIFIARISETHEYSPEAKAGAINSTMEALLMYFA
MILIARFRIENDSABHORMEANDTHEYWHOMILOVEDARETURNED
AGAINSTMEMYBONECLEAVETHTOMYSKINANDTOMYFLESHAND
IAMESCAPEDWITHTHESKINOFMYTEETHHAVEPITYUPONMEHA
VEPITYUPONMEOYEMYFRIENDSFORTHEHANDOFGODHATHTOU
CHEDMEWHYDOYEPERSECUTEMEASGODANDARENOTSATISFIE
DWITHMYFLESHOHTHATMYWORDSWERENOWWRITTEN!OHTHAT
THEYWEREINSCRIBEDINABOOK!THATWITHANIRONPENANDL
EADTHEYWEREGRAVENINTHEROCKFOREVER!BUTASFORMEIK
NOWTHATMYREDEEMERLIVETHANDATLASTHEWILLSTANDUPU
PONTHEEARTHANDAFTERMYSKINEVENTHISBODYISDESTROY
EDTHENWITHOUTMYFLESHSHALLISEEGODWHOMIEVENISHAL
LSEEONMYSIDEANDMINEEYESSHALLBEHOLDANDNOTASASTR
ANGERMYHEARTISCONSUMEDWITHINMEIFYESAYHOWWEWILL
PERSECUTEHIM!ANDTHATTHEROOTOFTHEMATTERISFOUNDI
NMEBEYEAFRAIDOFTHESWORDFORWRATHBRINGETHTHEPUNI
SHMENTSOFTHESWORDTHATYEMAYKNOWTHEREISAJUDGMENT
THENANSWEREDZOPHARTHENAAMATHITEANDSAIDTHEREFOR
EDOMYTHOUGHTSGIVEANSWERTOMEEVENBYREASONOFMYHAS
TETHATISINMEIHAVEHEARDTHEREPROOFWHICHPUTTETHME
TOSHAMEANDTHESPIRITOFMYUNDERSTANDINGANSWERETHM
EKNOWESTTHOUNOTTHISOFOLDTIMESINCEMANWASPLACEDU
PONEARTHTHATTHETRIUMPHINGOFTHEWICKEDISSHORTAND
THEJOYOFTHEGODLESSBUTFORAMOMENTTHOUGHHISHEIGHT
MOUNTUPTOTHEHEAVENSANDHISHEADREACHUNTOTHECLOUD
```

Verses in Matrix

The Book of Job 19:17-20:6

In order to differentiate Diana and Dodi from others on the globe with the same name, I needed to narrow the field, by placing them both in the Alma Tunnel as I conducted my searches.

Mohamed Al Fayed's first memorial, unveiled on 12th April 1998, consists of photos of the pair behind a pyramid-shaped display that holds a wine glass still smudged with lipstick from Diana's last dinner as well as a white and yellow gold, emerald and diamond engagement ring Dodi purchased just before they died, one which Mohamed believed Dodi was going to present to her on the night of their fatal accident.

Quote: Stephen Bates, The Guardian, Friday 19th October 2007. Dodi Fayed, the son of the Harrods owner Mohamed Al Fayed, did buy Princess Diana an £11,600 engagement ring a few hours before the couple's death in the Paris crash in 1997, though he had no chance to give it to her, the inquest jury heard yesterday.

The jury, sitting in the high court, was shown grainy CCTV footage from the Repossi jeweller's shop, across the square from the Ritz hotel in Paris, taken the afternoon before the crash. It showed Dodi Fayed, who was not accompanied by the princess, surrounded by staff as he inspected rings before leaving with a brochure.

Later footage showed Claude Roulet, assistant to the president of the Ritz hotel, visiting the shop and taking an item in a bag to the couple's room in the hotel's Imperial suite. The bag was deposited in the hotel strongroom before the couple went to Dodi's flat elsewhere in the city.

A ring bearing the inscription "Dis-moi Oui" ("Tell me Yes") was recovered from the flat after the couple's death. A receipt dated August 30th 1997, the day before the crash, listing a "bague de fiançaille" (engagement ring), was also recovered.

It has been a key part of Mohamed Fayed's case that the couple were about to get engaged, that the princess was pregnant and that they were murdered by British secret service agents on the instructions of the Duke of Edinburgh to prevent their marriage.

Mr Fayed says that the couple had met the jeweller Alberto Repossi at his Monte Carlo store to choose the ring during their holiday on the Fayed yacht in the Mediterranean. He claims that the ring was then sized in Italy and sent to Paris. There was no indication yesterday that Dodi Fayed had got round to giving Diana the ring but the jury was told that his father believes he was planning to present her with it on the night they died. End quote.

The engagement ring Dodi purchased for Diana on the day before they died, 30th August, 1997.

REPOSSI
JOAILLIERS
N° 01554

M. Monsieur Roulet (Hotel Ritz)

Adresse

Paris, le 30.08.97

		FRANCS F.	DEVISES

Confié — N° du passeport

Client nouveau ou ancien — Livraison

Lu et approuvé

OR DE VENDÔME - S.A.R.L. AU CAPITAL DE 1.500.000 FRANCS
6, Place Vendôme - 75001 PARIS - Tél. 01 42 96 42 34 - Fax 01 40 15 07 89

The receipt from Repossi Jewellers located at 6, Place Vendome, Paris, France, for the engagement ring Dodi purchased for Diana, showing the date, 30th August, 1997. Diana and Dodi died the following day, on 31st August, 1997.

References: http://dianaremembered.wordpress.com/2007/10/19/was-diana-to-wed-jury-see-receipt-for-engagement-ring-from-dodi/

http://www.telegraph.co.uk/news/worldnews/europe/france/1511268/Letter-from-Mohammed-Fayed-to-the-Intelligence-and-Security-Committee.html

The receipt clearly shows Dodi was about to propose: "bague de fiançaille" is French for engagement ring. Mohamed Al Fayed claims Dodi had spoken to Diana about the engagement and she had chosen the ring from two taken for her approval and selection as the receipt above shows. One ring has been rejected, and one selected.

Quote: Excerpt from Letter from Mohammed Fayed to the Intelligence and Security Committee.

12:01AM GMT 23 Feb 2006
Right Honourable Paul Murphy MP
Intelligence and Security Committee
70 Whitehall
London SW1A 2AS
14 February 2006

Dear Mr Murphy,
I suppose I should be encouraged to read in the press last week that the Intelligence and Security Committee is to question Mr Blair about his knowledge of what have been described as "CIA Torture Flights". That is absolutely right and proper and I trust that your Committee's questioning of him will be thorough and zealous.

It is certain that no actions of importance taken in relation to security activities in this country occurs without the knowledge and ultimate approval of the Prime Minister and those around him. They should be held to account.

However, I continue to be appalled by the fact that similar rigorous investigation by your committee has not occurred in relation to the actions and involvement of the security services regarding the murder of Princess Diana and my son Dodi.

How can it be that the position of overseas terrorist suspects receives such immediate and painstaking investigation by your committee, but when I, on behalf of the British public, as well as myself, shout out loud for investigation into security services involvement in the murder of Princess Diana (the mother of the future King of this country) and Dodi, my pleas go unanswered for some 8 years.

You clearly have remit to investigate all the activities of British security services including where they are acting on foreign soil. You clearly have remit to question the Prime Minister regarding his knowledge of, and complicity in those activities. Indeed, you have an obligation to parliament, and hence to the people of this country, to oversee their activities.

As is apparent from the attached press cutting, an overwhelming majority of the British public believe that Princess Diana and Dodi were murdered and there was direct involvement of British security services. According to the survey carried out by the newspaper that published the article, the percentage holding that belief was 94% Even if you, or numbers of your committee, are in the minority 6% you must still have a duty and obligation to the British public at large to investigate this matter.

In 1998, former MI6 agent Richard Tomlinson told the investigating judge in Paris that there were three MI6 agents who had an involvement in the assassination of Dodi and Princess Diana and/or in the subsequent cover up. How can you possibly tolerate and accept that this information about the wrongdoing of Messrs Spedding, Spearman and Langman can remain in the public domain for almost 8 years without comment, inquiry or investigation by your committee?

There is no doubt whatsoever that Princess Diana's and Dodi's conversations, particularly telephone conversations, and their movements were being monitored by the security services. There is equally no doubt that Messrs Langman, Spearman and Spedding have all been directly implicated, acting, I am sure, directly to the orders of the Royal Family, the Prime Minister and his senior henchmen. I note that Mr Langman has now also been directly implicated in the unlawful detention and torture of various migrant workers in Greece which gives clear indication that he would not hesitate to act unlawfully in overseeing, instigating or covering up the murder of my beloved son.

There is equally no doubt in my mind that such a momentous and horrific action would have been directly sanctioned by the Prime Minister. I suppose I should be grateful for the fact that the Coroner, Michael Burgess, asked the Commissioner of the Metropolitan Police to appoint officers to assist him; Lord Stevens then, of his own volition, decided that the allegations, concerns and complaints which had long been made by me, and which, until that time, had been largely ignored warranted full and thorough investigation.

He appointed a team of senior officers under his personal direction, the remit being conspiracy to murder. I suspect that, at the time, no one in authority ever thought that Lord Stevens and his team would find that I was justified in demanding answers to a number of critical questions.

However, although Lord Stevens has clearly been given some access to security service personnel and files, he has clearly not been given the truth, nor has he been given anything like the degree of access to which your committee are entitled.

Surely the death of a British Princess and my son warrants and deserves at least equal treatment and equal concern to foreign terrorist suspects; the overwhelming majority of people in this country would probably say that they deserve far more.

Against that background I urge and request you to demand information and answers from the security services and from Mr Blair to the many questions relating to the events of that night which they have so far successfully evaded.

We all know that there are those in power, including in the security services, who believe that the will of the people can safely be ignored in the cause of ensuring that the monarchy and the Establishment can do whatever they want, however horrendous, without being accountable to the British public.

Your committee has responsibility for ensuring that democracy is not trampled underfoot by those gangsters in the Establishment and the security services who believe that justice, freedom and fair play count for nothing. They believe that they are above the law. That may have been the case in the 18th Century but is not acceptable today.

Your committee seems to be obsessed with the fight against foreign terrorists, whilst ignoring the terrorism and criminality of those in authority in this country who think that they have no accountability to anyone. I, and 94% of the people in this country look to you to get to the truth and expose the real facts about this dreadful crime. It is time for your committee to bring those involved to justice. You should show that your committee is accountable to the British public and demonstrate that you will not tolerate the official terrorism that at present exists within our midst.

Given your accountability to the public I am sure that you will understand my reasons for publishing this letter; I owe it to the many millions of people who have personally expressed their support.

Yours sincerely,
M. Al Fayed End of Quote

It has been alleged that Sir David Spedding, Head of MI6, was ordered to organize the murder of Diana, Princess of Wales and her friend Dodi Al Fayed. Richard David Spearman, Chief of Staff for Sir David Spedding, was given an assignment and moved to Paris two weeks prior to the murder of Diana and her friend Dodi Al Fayed. Nicholas John Andrew Langman, Principal Assistant to Richard Spearman, was also involved with Spedding in the murder.

Reference: http://www.inside-news.ch/Tomlinson/List/Liste_Alpha.htm

This picture of Dodi Fayed, taken at interview shortly before his death, seems a precursor to the frustration that would be experienced by his father, Mohamed, other family members and friends who loved him, in the wake of his death and that of the woman he loved, Princess Diana.

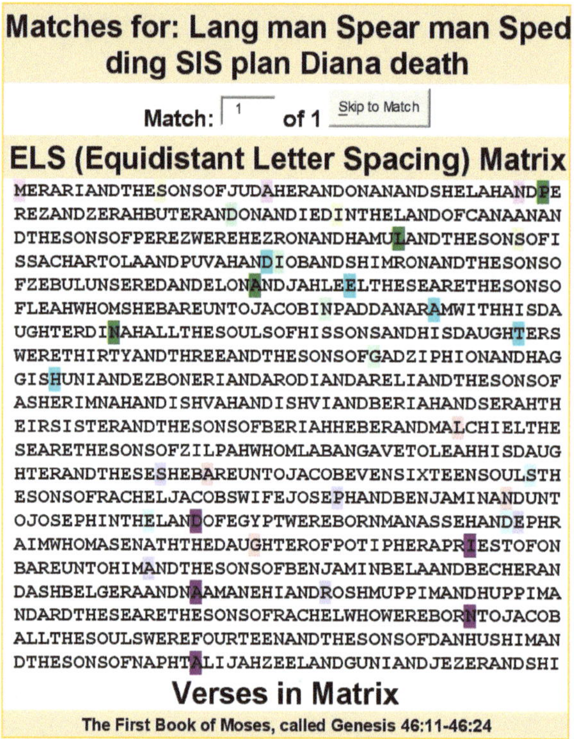

Matches for: Lang man Spear man Sped ding SIS plan Diana death

Match: 1 of 1 Skip to Match

ELS (Equidistant Letter Spacing) Matrix

MERARIANDTHESONSOFJUDAHERANDONANANDSHELAHANDPE
REZANDZERAHBUTERANDONANDIEDINTHELANDOFCANAANAN
DTHESONSOFPEREZWEREHEZRONANDHAMULANDTHESONSOFI
SSACHARTOLAANDPUVAHANDIOBANDSHIMRONANDTHESONSO
FZEBULUNSEREDANDELONANDJAHLEELTHESEARETHESONSO
FLEAHWHOMSHEBAREUNTOJACOBINPADDANARAMWITHHISDA
UGHTERDINAHALLTHESOULSOFHISSONSANDHISDAUGHTERS
WERETHIRTYANDTHREEANDTHESONSOFGADZIPHIONANDHAG
GISHUNIANDEZBONERIANDARODIANDARELIANDTHESONSOF
ASHERIMNAHANDISHVAHANDISHVIANDBERIAHANDSERAHTH
EIRSISTERANDTHESONSOFBERIAHHEBERANDMALCHIELTHE
SEARETHESONSOFZILPAHWHOMLABANGAVETOLEAHHISDAUG
HTERANDTHESESHEBAREUNTOJACOBEVENSIXTEENSOULSTH
ESONSOFRACHELJACOBSWIFEJOSEPHANDBENJAMINANDUNT
OJOSEPHINTHELANDOFEGYPTWEREBORNMANASSEHANDEPHR
AIMWHOMASENATHTHEDAUGHTEROFPOTIPHERAPRIESTOFON
BAREUNTOHIMANDTHESONSOFBENJAMINBELAANDBECHERAN
DASHBELGERAANDNAAMANEHIANDROSHMUPPIMANDHUPPIMA
NDARDTHESEARETHESONSOFRACHELWHOWEREBORNTOJACOB
ALLTHESOULSWEREFOURTEENANDTHESONSOFDANHUSHIMAN
DTHESONSOFNAPHTALIJAHZEELANDGUNIANDJEZERANDSHI

Verses in Matrix

The First Book of Moses, called Genesis 46:11-46:24

Section 4. How their much loved Dodi's death devastated his father and all his family. Dodi's Memorial at Oxted.

Reference: http://www.alfayed.com/dodi-and-diana/memorials.aspx

Set in the grounds of Mohamed Al Fayed's country estate in Oxted, Surrey, lies the tomb of Dodi. It was intended to be a temporary structure. The home was made for Dodi's grave to last until a more formal one would be erected out of Italian marble. When Mohamed Al Fayed saw the structure, he decided that it was so peaceful and fitting to Dodi's presence that it would remain forever. The structure is rugged, yet the atmosphere intensely dignified, with the echoing simplicity of the inscription, "Dodi".

There is nothing imposing about this remembrance site. There is a timber roof decorated with ivy, twelve supporting pillars, and a stone floor. The timber structure has become part of the landscape, open to the forces of nature. Designer Bill Mitchell: "There is some shelter, but it does not keep out the rain or the wind. It means Dodi is a part of nature."

Dodi's father says, "I wanted Dodi's final resting place to be somewhere beautiful and peaceful. In Egypt, we believe that the soul returns to the place that gave happiness, comfort and security. We feel that the soul continues to live in the place he loved and enjoyed. This was Dodi's favourite place, he grew up here from the age of ten and I feel his presence here.

I come here every day, often when I am tired, perhaps for two or three hours, and memories come back to me as I sit. I say prayers and think of Dodi, but I also sometimes do my work here, or take breakfast. My children come here and play around and run. There is a nice atmosphere."

Memorial to Dodi and Diana at Harrods, "The Innocents", erected by Dodi's father, Mohamed Al Fayed.

Mohamed also claims a nurse at the hospital where Diana was taken, (who asked that her name be shielded, and in fact, kept her hand over her name badge as she spoke to him), said that Diana had given a request before her death for her older sister Sarah to take care of her boys. Sarah was Charles' former girlfriend, before he became engaged to marry Diana. However, the exact wording of her statement I have not been able to find, just the general gist of the message the nurse allegedly relayed to Mohamed from Diana.

Matches for: Diana Fayed gold Dis moi Oui ring too big

Match: 1 **of 2** Skip to Match

ELS (Equidistant Letter Spacing) Matrix

```
THOUSANDSEVENHUNDREDANDFIFTYSHEKELSFORTHEMENOF
WARHADTAKENBOOTYEVERYMANFORHIMSELFANDMOSESANDE
LEAZARTHEPRIESTTOOKTHEGOLDOFTHECAPTAINSOFTHOUS
ANDSANDOFHUNDREDSANDBROUGHTITINTOTHETENTOFMEET
INGFORAMEMORIALFORTHECHILDRENOFISRAELBEFOREJEH
OVAHNOWTHECHILDRENOFREUBENANDTHECHILDRENOFGADH
ADAVERYGREATMULTITUDEOFCATTLEANDWHENTHEYSAWTHE
LANDOFJAZERANDTHELANDOFGILEADTHATBEHOLDTHEPLAC
EWASAPLACEFORCATTLETHECHILDRENOFGADANDTHECHILD
RENOFREUBENCAMEANDSPAKEUNTOMOSESANDTOELEAZARTH
EPRIESTANDUNTOTHEPRINCESOFTHECONGREGATIONSAYIN
GATAROTHANDDIBONANDJAZERANDNIMRAHANDHESHBONAND
ELEALEHANDSEBAMANDNEBOANDBEONTHELANDWHICHJEHOV
AHSMOTEBEFORETHECONGREGATIONOFISRAELISALANDFOR
CATTLEANDTHYSERVANTSHAVECATTLEANDTHEYSAIDIFWEH
AVEFOUNDFAVORINTHYSIGHTLETTHISLANDBEGIVENUNTOT
HYSERVANTSFORAPOSSESSIONBRINGUSNOTOVERTHEJORDA
NANDMOSESSAIDUNTOTHECHILDRENOFGADANDTOTHECHILD
RENOFREUBENSHALLYOURBRETHRENGOTOTHEWARANDSHALL
YESITHEREANDWHEREFOREDISCOURAGEYETHEHEARTOFTHE
CHILDRENOFISRAELFROMGOINGOVERINTOTHELANDWHICHJ
EHOVAHHATHGIVENTHEMTHUSDIDYOURFATHERSWHENISENT
THEMFROMKADESHBARNEATOSEETHELANDFORWHENTHEYWEN
TUPUNTOTHEVALLEYOFESHCOLANDSAWTHELANDTHEYDISCO
URAGEDTHEHEARTOFTHECHILDRENOFISRAELTHATTHEYSHO
ULDNOTGOINTOTHELANDWHICHJEHOVAHHADGIVENTHEMAND
```

Verses in Matrix

The Fourth Book of Moses, called Numbers 31:52-32:9

One of the difficulties for anyone researching The Bible Code, is believing any results that include the ordinary text, that is the linear text, of The Bible. The above matrix has the words of the ordinary text "ring" and "gold" intersecting with words of the matrix: "Diana" with "ring", and "too big", "Dis moi Oui" (Tell me Yes), with "gold".

To be quite honest, I found this fascinating, and it left me wondering if Rips' and Drosnin's beliefs could be true: that words of the ordinary Biblical text, (especially if they cross or intersect the "crossword"), carry weight in the final analysis of the matrix veracity.

Mohamed Al Fayed's claims continued that, in consideration of a home to share together, the couple had looked at the Windsor Villa in the Bois de Boulogne, France, restored by Dodi's father. This was formerly the last residence owned by Edward VIII, who abdicated/renounced the English throne to marry the thrice-divorced Mrs. Wallis Simpson.

For Diana and Dodi to have lived their married lives together at the site of the once disgraced and exiled royals might have been a bitter pill for royalty to swallow perhaps? If it was true that Diana was pregnant, the ring might not have been an engagement ring, as the jeweller's receipt showed, but a wedding ring.

After all, it is not the kind of thing a couple confides in their jeweller: that the intended bride is pregnant, especially when she is someone of Diana's stature, constantly under public scrutiny and gaze. If Diana feared for her safety, all the more reason why she would want to keep secret, if possible, her pregnancy and forthcoming marriage.

Section 5. Diana's sister Lady Sarah McCorquodale and her mother Lady Frances Shand Kydd, without obtaining permission to do so, shredded evidence that may have been valuable to a court of inquiry into Diana's and Dodi's deaths and to an accurate historical account of Diana's life as Princess of Wales, (former) wife of Charles and mother of William, heirs to the throne.

By Richard Alleyne
7:00PM GMT 28 Jan 2008

http://www.telegraph.co.uk/news/uknews/1576767/Diana-thought-Fayed-was-spying-on-her.html

Diana, Princess of Wales, thought Mohamed Fayed was spying on her during her final stay on his luxury yacht, her sister told the inquest into her death has heard. The Princess was said to have revealed her suspicions to her sister Lady Sarah McCorquodale in a telephone call just days before she was killed in a Paris car crash. She was calling from the Fayed yacht, The Jonikal, where she was on a Mediterranean cruise with Dodi Fayed, son of the Harrods tycoon in August 1997, the hearing was told. Ian Burnett QC, for the coroner, asked: "Did the Princess speak to you from the boat about bugging?"

"Yes, she thought the boat was being bugged by Mr Fayed senior," Lady Sarah told the central London inquest. She recalled that the Princess had called her and she "assumed" it was on her mobile telephone.

Mr Fayed believes the Princess was killed by MI6 on the orders of the Duke of Edinburgh because she was pregnant with Dodi's child and the couple were about to get engaged. But from the evidence of her last telephone call with her sister, just two days before she died, Lady Sarah believed that the romance with Dodi was on its last legs. The court has also heard the Princess feared she might be bugged by members of the security services and also believed predictions by astrologers that somebody was out to get her.

Lady Sarah recalled her sister was "distraught" because she felt she had been negatively misquoted in a French newspaper about her landmines campaign. Lady Sarah told the court: "There had been an article in a French newspaper Le Monde about landmines."

"She was very upset because she felt she had been misquoted and as a result had appeared to look like she was criticising the Government. She was distraught. Lady Sarah suggested her sister speak to Dodi about it only for the Princess to reply "that would be a waste of time". She said: "From that, I just did not think that the relationship had much longer to go."

Lady Sarah was questioned about her sister's relationship with heart surgeon Hasnat Khan, and revealed she thought the couple might have wed. "I believe there was a strong possibility that they might have married," she told the court. Asked about how the couple's two-year relationship ended in the summer of 1997 she replied: "He couldn't take, if that's the right word, her lifestyle and was a very committed surgeon and the two didn't go together. I don't think she believed the relationship was ended or she hoped it wasn't."

Lady Sarah and her mother Frances Shand Kydd, who died in 2004, were executors of the Princess's estate and helped to sort out her personal belongings at Kensington Palace. The inquest has already heard evidence how Mrs Shand Kydd spent days shredding documents found in her daughter's home. Lady Sarah told the hearing: "We agreed beforehand that anything that was sensitive or would be distressing to the boys then, or in 30 years time, should be destroyed and that's what we both did because there was a lot of correspondence." She said they only destroyed things like "thank you" notes but "nothing historical was ever shredded", and she agreed with Mr Burnett's suggestion that no letters from the Royal family were binned.

Lady Sarah was also asked about a statement her sister made to the press while in St Tropez in July 1997 when she told them "you will get a big surprise with the next thing I do". The Princess's eldest sister told the court: "Every time she took a battering from the press she talked about (moving to) France, America and then latterly South Africa where my brother was living."

Mr Fayed is convinced that his son Dodi had bought an engagement ring and was planning to ask Diana to marry him. But the inquest heard the Princess did not tell her sister about any gifts she had received from her lover. Mr Burnett QC also asked Lady Sarah if there was any mention by the Princess of an engagement, or being pregnant - another suggestion maintained by Mr Fayed. She replied "none" to both questions. Questioned about who, apart from friends, the Princess might have discussed the issues of marriage and having a child with, Lady Sarah said: "I believe she would have talked to William her son. They were very close." End of Quote.

What Sarah and Frances did was without doubt a great loss both to the court and to the nation. It was not up to them to decide what was or was not valuable evidence or of historical worth. If they were shredding things they thought "might distress William and Harry, either then or in 30 years time", as they stated, it rather sounds like sensitive documentation was destroyed, and it is therefore eminently possible they have concealed (by the destruction of it), evidence of importance to the court.

"Thank you notes" would not have distressed anybody, yet even they could have contained vitally important evidence to Diana's case. We have to wonder at their motives for doing this so stealthily, without obtaining permission for their course of action. Who were they really protecting? Themselves, and the Spencer family name? Or like Simone Simmons, did they take this drastic course of (what they must have known was quite possibly illegal and therefore prosecutable) action, out of morbid fear for their own or others' lives?

Sarah, Diana's sister left, and their mother Frances Shand-Kydd.

There were two Sarahs close to Diana. Sarah Ferguson, "Fergie" Prince Andrew's wife, and her sister Sarah. Is it possible that her dying wishes may have been misinterpreted? It is not without the bounds of possibility that Diana might have requested Sarah Ferguson, her great friend and sister-in-law, to take care of her boys. After all, she would have known her sons of necessity, as royal heirs to the throne, would be ensconced within the Royal Family after her death. Sarah Ferguson was divorced from Andrew in 1996, just prior to Diana's death.

However, given that they were both divorcees with royal children, she might have thought it wise for Sarah Ferguson to keep a lookout for the children's welfare. She would have known "Fergie" would continue to move within the Royal Family circle, especially with Andrew, Charles' brother. It must not be forgotten that Andrew was Diana's childhood playmate, and that William and Harry were then and still are, close to Beatrice and Eugenie, who are fifth and sixth in line of succession to the throne.

Some reports say that Fergie and Diana had fallen out just prior to Diana's death, over a flippant remark that Fergie made to the press – that she had contracted a plantar wart from wearing a pair of Diana's shoes.

It is not possible to input a word with only two letters into the matrix. Unless the unbroken word "sister-in-law" comes up, I can only input "sister" and "law". Of course, the Diana and Sarah mentioned here could be any two people on the globe with the same name, so it is not a definitive finding.

Field of flowers at Kensington Palace, many for both Diana and Dodi.

Matches for: Diana want Sarah sister law take care her boys

Match: 1 of 1 Skip to Match

ELS (Equidistant Letter Spacing) Matrix

HIRELINGTHEGLORYOFMOABSHALLBEBROUGHTINTOCONTEM
PTWITHALLHISGREATMULTITUDEANDTHEREMNANTSHALLBE
VERYSMALLANDOFNOACCOUNTTHEBURDENOFDAMASCUSBEHO
LDDAMASCUSISTAKENAWAYFROMBEINGACITYANDITSHALLB
EARUINOUSHEAPTHECITIESOFAROERAREFORSAKENTHEYSH
ALLBEFORFLOCKSWHICHSHALLLIEDOWNANDNONESHALLMAK
ETHEMAFRAIDANDTHEFORTRESSSHALLCEASEFROMEPHRAIM
ANDTHEKINGDOMFROMDAMASCUSANDTHEREMNANTOFSYRIAT
HEYSHALLBEASTHEGLORYOFTHECHILDRENOFISRAELSAITH
JEHOVAHOFHOSTSANDITSHALLCOMETOPASSINTHATDAYTHA
TTHEGLORYOFJACOBSHALLBEMADETHINANDTHEFATNESSOF
HISFLESHSHALLWAXLEANANDITSHALLBEASWHENTHEHARVE
STMANGATHERETHTHESTANDINGGRAINANDHISARMREAPETH
THEEARSYEAITSHALLBEASWHENONEGLEANETHEARSINTHEV
ALLEYOFREPHAIMYETTHERESHALLBELEFTTHEREINGLEANI
NGSASTHESHAKINGOFANOLIVETREETWOORTHREEBERRIESI
NTHETOPOFTHEUPPERMOSTBOUGHFOURORFIVEINTHEOUTMO
STBRANCHESOFAFRUITFULTREESAITHJEHOVAHTHEGODOFI
SRAELINTHATDAYSHALLMENLOOKUNTOTHEIRMAKERANDTHE
IREYESSHALLHAVERESPECTTOTHEHOLYONEOFISRAELANDT

Verses in Matrix

The Book of the Prophet Isaiah 16:14-17:6

Matches for: Diana dies Alma Tunnel Fiat hits Merc

Match: 1 of 2 Skip to Match

ELS (Equidistant Letter Spacing) Matrix

HATHSWORNJEHOVAHWILLHAVEWARWITHAMALEKFROMGENER
ATIONTOGENERATIONNOWJETHROTHEPRIESTOFMIDIANMOS
ESFATHERINLAWHEARDOFALLTHATGODHADDONEFORMOSESA
NDFORISRAELHISPEOPLEHOWTHATJEHOVAHHADBROUGHTIS
RAELOUTOFEGYPTANDJETHROMOSESFATHERINLAWTOOKZIP
PORAHMOSESWIFEAFTERHEHADSENTHERAWAYANDHERTWOSO
NSOFWHOMTHENAMEOFTHEONEWASGERSHOMFORHESAIDIHAV
EBEENASOJOURNERINAFOREIGNLANDANDTHENAMEOFTHEOT
HERWASELIEZERFORHESAIDTHEGODOFMYFATHERWASMYHEL
PANDDELIVEREDMEFROMTHESWORDOFPHARAOHANDJETHROM
OSESFATHERINLAWCAMEWITHHISSONSANDHISWIFEUNTOMO
SESINTOTHEWILDERNESSWHEREHEWASENCAMPEDATTHEMOU
NTOFGODANDHESAIDUNTOMOSESITHYFATHERINLAWJETHRO
AMCOMEUNTOTHEEANDTHYWIFEANDHERTWOSONSWITHHERAN
DMOSESWENTOUTTOMEETHISFATHERINLAWANDDIDOBEISAN
CEANDKISSEDHIMANDTHEYASKEDEACHOTHEROFTHEIRWELF
AREANDTHEYCAMEINTOTHETENTANDMOSESTOLDHISFATHER
INLAWALLTHATJEHOVAHHADDONEUNTOPHARAOHANDTOTHEE
GYPTIANSFORISRAELSSAKEALLTHETRAVAILTHATHADCOME
UPONTHEMBYTHEWAYANDHOWJEHOVAHDELIVEREDTHEMANDJ
ETHROREJOICEDFORALLTHEGOODNESSWHICHJEHOVAHHADD

Verses in Matrix

The Second Book of Moses, Called Exodus 17:16-18:9

Matches for: Fayed wed ring Italy sent suite die soon after

Match: 1 of 1 Skip to Match

ELS (Equidistant Letter Spacing) Matrix

```
DALLDRINKTHATMAYBEDRUNKINEVERYSUCHVESSELSHALLB
EUNCLEANANDEVERYTHINGWHEREUPONANYPARTOFTHEIRCA
RCASSFALLETHSHALLBEUNCLEANWHETHEROVENORRANGEFO
RPOTSITSHALLBEBROKENINPIECESTHEYAREUNCLEANANDS
HALLBEUNCLEANUNTOYOUNEVERTHELESSAFOUNTAINORAPI
TWHEREINISAGATHERINGOFWATERSHALLBECLEANBUTTHAT
WHICHTOUCHETHTHEIRCARCASSSHALLBEUNCLEANANDIFAU
GHTOFTHEIRCARCASSFALLUPONANYSOWINGSEEDWHICHIST
OBESOWNITISCLEANBUTIFWATERBEPUTUPONTHESEEDANDA
UGHTOFTHEIRCARCASSFALLTHEREONITISUNCLEANUNTOYO
UANDIFANYBEASTOFWHICHYEMAYEATDIEHETHATTOUCHETH
THECARCASSTHEREOFSHALLBEUNCLEANUNTILTHEEVENAND
HETHATEATETHOFTHECARCASSOFITSHALLWASHHISCLOTHE
SANDBEUNCLEANUNTILTHEEVENHEALSOTHATBEARETHTHEC
ARCASSOFITSHALLWASHHISCLOTHESANDBEUNCLEANUNTIL
THEEVENANDEVERYCREEPINGTHINGTHATCREEPETHUPONTH
EEARTHISANABOMINATIONITSHALLNOTBEEATENWHATSOEV
ERGOETHUPONTHEBELLYANDWHATSOEVERGOETHUPONALLFO
URSORWHATSOEVERHATHMANYFEETEVENALLCREEPINGTHIN
GSTHATCREEPUPONTHEEARTHTHEMYESHALLNOTEATFORTHE
```

Verses in Matrix

The Third Book of Moses, commonly called Leviticu 11:34-11:40

Section 6. Diana was pregnant when she died: French police.

And what of the claim made by Mohamed Al Fayed and others that Princess Diana was pregnant when she died? Quote: Last updated at: (Beijing Time) Tuesday, December 23rd, 2003. Princess Diana was pregnant when she died in a Paris car crash in 1997 with her lover Dodi Fayed at her side, a "senior French police source" told the Independent on Sunday. "I can tell you that she was pregnant," the broadsheet quoted the source as saying Sunday. End of Quote.

References:
http://english.peopledaily.com.cn/200312/23/eng20031223_131042.shtml

2. Broadsheet is the largest of the various newspaper formats and is characterized by long vertical pages (typically 22 inches / 559 millimetres or more).

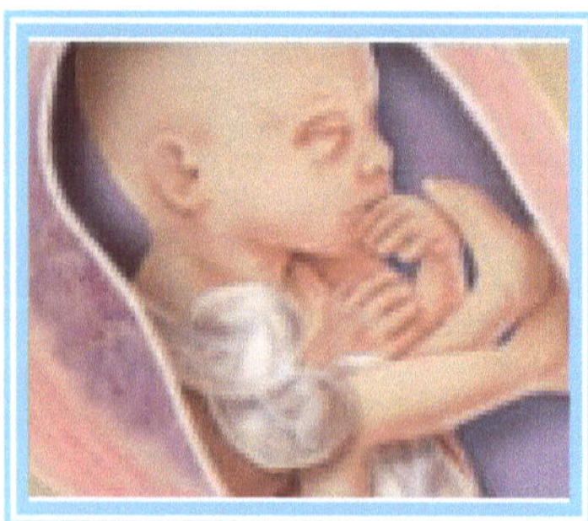

Matches for: Diana dead two month boy child die Alma Tunnel

Match: 1 of 1 Skip to Match

ELS (Equidistant Letter Spacing) Matrix

```
DTHEYTURNEDANDFLEDTOWARDTHEWILDERNESSUNTOTHERO
CKOFRIMMONANDTHEYGLEANEDOFTHEMINTHEHIGHWAYSFIV
ETHOUSANDMENANDFOLLOWEDHARDAFTERTHEMUNTOGIDOMA
NDSMOTEOFTHEMTWOTHOUSANDMENSOTHATALLWHOFELLTHA
TDAYOFBENJAMINWERETWENTYANDFIVETHOUSANDMENTHAT
DREWTHESWORDALLTHESEWEREMENOFVALORBUTSIXHUNDRE
DMENTURNEDANDFLEDTOWARDTHEWILDERNESSUNTOTHEROC
KOFRIMMONANDABODEINTHEROCKOFRIMMONFOURMONTHSAN
DTHEMENOFISRAELTURNEDAGAINUPONTHECHILDRENOFBEN
JAMINANDSMOTETHEMWITHTHEEDGEOFTHESWORDBOTHTHEE
NTIRECITYANDTHECATTLEANDALLTHATTHEYFOUNDMOREOV
ERALLTHECITIESWHICHTHEYFOUNDTHEYSETONFIRENOWTH
EMENOFISRAELHADSWORNINMIZPAHSAYINGTHERESHALLNO
TANYOFUSGIVEHISDAUGHTERUNTOBENJAMINTOWIFEANDTH
EPEOPLECAMETOBETHELANDSATTHERETILLEVENBEFOREGO
DANDLIFTEDUPTHEIRVOICESANDWEPTSOREANDTHEYSAIDO
JEHOVAHTHEGODOFISRAELWHYISTHISCOMETOPASSINISRA
ELTHATTHERESHOULDBETODAYONETRIBELACKINGINISRAE
LANDITCAMETOPASSONTHEMORROWTHATTHEPEOPLEROSEEA
RLYANDBUILTTHEREANALTARANDOFFEREDBURNTOFFERING
SANDPEACEOFFERINGSANDTHECHILDRENOFISRAELSAIDWH
OISTHEREAMONGALLTHETRIBESOFISRAELTHATCAMENOTUP
INTHEASSEMBLYUNTOJEHOVAHFORTHEYHADMADEAGREATOA
THCONCERNINGHIMTHATCAMENOTUPUNTOJEHOVAHTOMIZPA
```

Verses in Matrix

The Book of Judges 20:45-21:5

When doing the search above, "one" "two" "four" and "five" came up in the same matrix. Reportedly Diana had told a medic before she died that she was six weeks pregnant. If true, that would mean she was in her second month of pregnancy. I therefore have included only the matrix containing the word "two".

The sons of Prince Charles and Princess Diana. William marries Kate, while Harry is Best Man. The wedding of joy was for many, especially the royal family, no doubt, a welcome relief from the distress of the ongoing French court investigations into their mother's and Dodi Fayed's deaths.

Matches for: Diana dies with child Alma Tunnel two son's Mum

Match: 1 **of 1** Skip to Match

ELS (Equidistant Letter Spacing) Matrix

```
SSUNTOTHEROCKOFRIMMONANDTHEYGLEANEDOFTHEMINTHE
HIGHWAYSFIVETHOUSANDMENANDFOLLOWEDHARDAFTERTHE
MUNTOGIDOMANDSMOTEOFTHEMTWOTHOUSANDMENSOTHATAL
LWHOFELLTHATDAYOFBENJAMINWERETWENTYANDFIVETHOU
SANDMENTHATDREWTHESWORDALLTHESEWEREMENOFVALORB
UTSIXHUNDREDMENTURNEDANDFLEDTOWARDTHEWILDERNES
SUNTOTHEROCKOFRIMMONANDABODEINTHEROCKOFRIMMONF
OURMONTHSANDTHEMENOFISRAELTURNEDAGAINUPONTHECH
ILDRENOFBENJAMINANDSMOTETHEMWITHTHEEDGEOFTHESW
ORDBOTHTHEENTIRECITYANDTHECATTLEANDALLTHATTHEY
FOUNDMOREOVERALLTHECITIESWHICHTHEYFOUNDTHEYSET
ONFIRENOWTHEMENOFISRAELHADSWORNINMIZPAHSAYINGT
HERESHALLNOTANYOFUSGIVEHISDAUGHTERUNTOBENJAMIN
TOWIFEANDTHEPEOPLECAMETOBETHELANDSATTHERETILLE
VENBEFOREGODANDLIFTEDUPTHEIRVOICESANDWEPTSOREA
NDTHEYSAIDOJEHOVAHTHEGODOFISRAELWHYISTHISCOMET
OPASSINISRAELTHATTHERESHOULDBETODAYONETRIBELAC
KINGINISRAELANDITCAMETOPASSONTHEMORROWTHATTHEP
EOPLEROSEEARLYANDBUILTTHEREANALTARANDOFFEREDBU
RNTOFFERINGSANDPEACEOFFERINGSANDTHECHILDRENOFI
SRAELSAIDWHOISTHEREAMONGALLTHETRIBESOFISRAELTH
ATCAMENOTUPINTHEASSEMBLYUNTOJEHOVAHFORTHEYHADM
ADEAGREATOATHCONCERNINGHIMTHATCAMENOTUPUNTOJEH
OVAHTOMIZPAHSAYINGHESHALLSURELYBEPUTTODEATHAND
```

Verses in Matrix

The Book of Judges 20:45-21:5

As you can see from the above Matrix, the words "children" and "child" in the ordinary text cross "Mum", so may be considered not legitimate search results. However, Rips and Drosnin may consider that crossing significant, giving weight to the search results.

I tried inserting the negation of the above, using the words "not true" and "false" but they did not come up. If The Bible Code is true, then it may be true that Diana was indeed pregnant at the time of the accident.

If we look at the Matrix below, we can see that it may be considered ambiguous. The word "lies" might mean that the whole thing is a lie, or it might mean that when Diana lay dead, she had a boy child in her womb, who would have been half-brother to William and Harry.

If the Bible Code is true, then regardless of whether or not the child Diana carried belonged to Dodi, the tragedy of her death in a car smash is worsened by the fact that a pregnant mother has died, her unborn child with her. If a murder has been committed, it is all the more heinous if the assassins knew she was a mother-to-be. Such cruelty against an innocent unborn child (and its parent or parents) renders the crime all the more unconscionable.

Matches for: Diana dies Alma Tunnel with boy child lies dead

Match: 1 of 1 Skip to Match

ELS (Equidistant Letter Spacing) Matrix

```
DERNESSUNTOTHEROCKOFRIMMONANDTHEYGLEANEDOFTHEM
INTHEHIGHWAYSFIVETHOUSANDMENANDFOLLOWEDHARDAFT
ERTHEMUNTOGIDOMANDSMOTEOFTHEMTWOTHOUSANDMENSOT
HATALLWHOFELLTHATDAYOFBENJAMINWERETWENTYANDFIV
ETHOUSANDMENTHATDREWTHESWORDALLTHESEWEREMENOFV
ALORBUTSIXHUNDREDMENTURNEDANDFLEDTOWARDTHEWILD
ERNESSUNTOTHEROCKOFRIMMONANDABODEINTHEROCKOFRI
MMONFOURMONTHSANDTHEMENOFISRAELTURNEDAGAINUPON
THECHILDRENOFBENJAMINANDSMOTETHEMWITHTHEEDGEOF
THESWORDBOTHTHEENTIRECITYANDTHECATTLEANDALLTHA
TTHEYFOUNDMOREOVERALLTHECITIESWHICHTHEYFOUNDTH
EYSETONFIRENOWTHEMENOFISRAELHADSWORNINMIZPAHSA
YINGTHERESHALLNOTANYOFUSGIVEHISDAUGHTERUNTOBEN
JAMINTOWIFEANDTHEPEOPLECAMETOBETHELANDSATTHERE
TILLEVENBEFOREGODANDLIFTEDUPTHEIRVOICESANDWEPT
SOREANDTHEYSAIDOJEHOVAHTHEGODOFISRAELWHYISTHIS
COMETOPASSINISRAELTHATTHERESHOULDBETODAYONETRI
BELACKINGINISRAELANDITCAMETOPASSONTHEMORROWTHA
TTHEPEOPLEROSEEARLYANDBUILTTHEREANALTARANDOFFE
REDBURNTOFFERINGSANDPEACEOFFERINGSANDTHECHILDR
ENOFISRAELSAIDWHOISTHEREAMONGALLTHETRIBESOFISR
AELTHATCAMENOTUPINTHEASSEMBLYUNTOJEHOVAHFORTHE
YHADMADEAGREATOATHCONCERNINGHIMTHATCAMENOTUPUN
TOJEHOVAHTOMIZPAHSAYINGHESHALLSURELYBEPUTTODEA
```

Verses in Matrix

The Book of Judges 20:45-21:5

The Brits have decided to let the matter lie, put the baby to bed as it were, considering there is no more to be gained from further rehashing old evidence. Yet, even if Britain has lain the inquiry to rest, as at January 2012, further investigation was continuing in France. Charges were made that evidence was withheld from the previous inquest in France by British police. A lot of people could be forgiven for thinking the esteemed British Scotland Yard constabulary had something to hide.

Tuesday May 10 2011 by Frank Thorne in Sydney Quote: http://www.express.co.uk/posts/view/245718

Diana Inquest Samples Switched. Princess Diana's post-mortem samples were switched with those from another woman, an explosive new book claims. Adding weight to "cover-up" theories, it says samples were swapped prior to toxicological testing. According to documents uncovered for the book, published this week, the toxicologist at London's Charing Cross Hospital received the samples of another female and tested them in the belief that they were from Princess Diana.

In his latest volume in a series about the Diana inquest, author John Morgan believes he has discovered the truth of what occurred in the 24 hours following the deaths of the Princess and Dodi Al Fayed in a Paris car crash on August 31, 1997.

"There is a lot of evidence which points to the toxicology testing being carried out on samples that did not come from the body of Princess Diana," he said last night, pointing out that the documents were, along with others, withheld from the inquest jury.

Mr Morgan said he had uncovered a litany of conflicting evidence, inconsistencies, mis-labelling of body samples, cover-ups, evidence and witnesses who were never called to give evidence at the inquest.

He is now calling for independent DNA tests to be carried out on the body samples. "The samples at Charing Cross Hospital have never been subjected to DNA testing. With so much conflicting evidence, how can we be sure?" he said. "The evidence I have studied indicates that there are two lots of samples. One belongs to Diana, which is held by the Metropolitan Police's Operation Paget, and the other lot are samples from another body and held by Charing Cross Hospital.

"Diana's UK post-mortem samples were switched ahead of the toxicology testing." He added: "The jury were not given the post mortem and toxicology reports on Diana. If they had, they should have been able to work out that the toxicology testing was conducted on samples that weren't Diana's.

"For example, Diana's body was embalmed in France, but there was no embalming fluid in the toxicology tested samples." He went on: "Diana had consumed alcohol that night in Dodi's apartment and later at the Ritz Hotel. It is recorded by the Hammersmith and Fulham mortuary manager that her stomach smelled strongly of alcohol, but there was no alcohol in the samples tested in London."

Mr Morgan said that within 24 hours of her death, Diana's body had been subjected to embalming in France and the UK, along with two post mortems.

"I am asking, "Why?"" said the Australian-based writer who lives in Brisbane and insisted that documentation shows interference from senior aides on behalf of the Queen.

"Having removed Princess Diana as a member of the Royal Family in 1996, suddenly, Diana became royal again. Only the Queen could have ordered this. The Queen took control of events from Balmoral, very early following Diana's death," Mr Morgan said, adding that he believed the UK coroner should not have taken possession of Diana's body after it arrived back in England.

By law, jurisdiction over the body should have gone to the coroner in Northamptonshire, which covers Diana's family home, Althorp, where it was known Diana's body would be buried."

Another stark example of a switch Mr Morgan claims to have discovered was that there was no vitreous humour (eyeball gel) sample taken during the UK post-mortem, yet a sample was tested by the London toxicologist.

Also, sample labels received by the toxicologist, Susan Paterson, for both blood and liver tests, read differently to the descriptions in the post-mortem documentation. Mr Morgan has spent more than five years investigating the evidence relating to the 1997 Paris crash in the city's Alma Tunnel.

His series of books is based on the testimony heard during the inquest and evidence in official documents from within the British police investigation that were withheld from the inquest jury.

Leading QC Michael Mansfield, who served throughout the six months of the London inquest, stated last year: "I have no doubt that the volumes written by him will come to be regarded as the Magnum Opus on the death crash.
Dodi's father, Mohamed Al Fayed, has praised Mr Morgan's work, calling it "heroic" and "impressive works of forensic enquiry and immensely helpful to the cause of truth."

Mr Al Fayed said: "I believe that John Morgan has done more to expose the facts of this case than the police in France and Britain."

Reference: The British Cover-Up by John Morgan, published by Lightning Source UK Ltd, available on Amazon UK, (www.amazon.co.uk.)

For information about Dr. Susan Paterson, Toxicologist, please see also: http://www1.imperial.ac.uk/departmentofmedicine/divisions/diabetesendocrinologymetabolism/meta/toxicology/

Charing Cross Hospital where Diana's samples were taken for testing.

Seconds before the fatal crash. It can clearly be seen that Trevor Rees Jones, the only survivor, has done up his seat belt. None of the other occupants of the car, Henri Paul, Dodi or Diana are wearing theirs.

Matches for: Diana died Alma Tunnel wrote note fear hurt head

Match: 1 **of 1** Skip to Match

ELS (Equidistant Letter Spacing) Matrix

```
HOVAHWILLHAVEWARWITHAMALEKFROMGENERATIONTOGENE
RATIONNOWJETHROTHEPRIESTOFMIDIANMOSESFATHERINL
AWHEARDOFALLTHATGODHADDONEFORMOSESANDFORISRAEL
HISPEOPLEHOWTHATJEHOVAHHADBROUGHTISRAELOUTOFEG
YPTANDJETHROMOSESFATHERINLAWTOOKZIPPORAHMOSESW
IFEAFTERHEHADSENTHERAWAYANDHERTWOSONSOFWHOMTHE
NAMEOFTHEONEWASGERSHOMFORHESAIDIHAVEBEENASOJOU
RNERINAFOREIGNLANDANDTHENAMEOFTHEOTHERWASELIEZ
ERFORHESAIDTHEGODOFMYFATHERWASMYHELPANDDELIVER
EDMEFROMTHESWORDOFPHARAOHANDJETHROMOSESFATHERI
NLAWCAMEWITHHISSONSANDHISWIFEUNTOMOSESINTOTHEW
ILDERNESSWHEREHEWASENCAMPEDATTHEMOUNTOFGODANDH
ESAIDUNTOMOSESITHYFATHERINLAWJETHROAMCOMEUNTOT
HEEANDTHYWIFEANDHERTWOSONSWITHHERANDMOSESWENTO
UTTOMEETHISFATHERINLAWANDDIDOBEISANCEANDKISSED
HIMANDTHEYASKEDEACHOTHEROFTHEIRWELFAREANDTHEYC
AMEINTOTHETENTANDMOSESTOLDHISFATHERINLAWALLTHA
TJEHOVAHHADDONEUNTOPHARAOHANDTOTHEEGYPTIANSFOR
ISRAELSSAKEALLTHETRAVAILTHATHADCOMEUPONTHEMBYT
HEWAYANDHOWJEHOVAHDELIVEREDTHEMANDJETHROREJOIC
```

Verses in Matrix

The Second Book of Moses, Called Exodus 17:16-18:8

Section 7. The missing Mischon Note. Lord Condon gives evidence to French inquiry over 'suppressed Diana evidence'.

Quote: 2:49AM GMT 26 Jan 2012 The Telegraph. 26th January, 2012. Lord Condon, a former Met Police chief has given evidence at a French inquiry into allegations that he suppressed evidence in the investigation into the death of Diana, Princess of Wales.

Lord Condon, who was commissioner from 1993 to 2000, and former assistant commissioner Sir David Veness said they had voluntarily offered to provide witness evidence to the French Juge d'Instruction.

Former Harrods boss Mohammed al Fayed, the father of Dodi, who also died in the August 1997 Paris car crash, has long accused the pair of deliberately withholding evidence from the French investigation into their deaths.

Lord Condon said: "Sir David and I volunteered to go to Paris as witnesses to help the French authorities explore and then bring to a conclusion these repeated complaints made by Mr Al Fayed. We trust that our evidence was helpful, and we both sincerely hope that for all concerned, including Mr al Fayed, some closure can be finally brought to this tragic case." End of Quote.

Tiggy Legge-Bourke and her ancestral home. Tiggy was nanny to William and Harry, and personal assistant to Prince Charles. Diana reportedly said that Tiggy had an abortion while in her relationship with Charles. She had assumed that Charles wanted to marry Tiggy. This assumption proved incorrect. He married his former youth sweetheart Camilla.

Matches for: Diana send note true not fake die Alma Tunnel

Match: 1 of 1 Skip to Match

ELS (Equidistant Letter Spacing) Matrix

```
NHISOWNHOUSEBUTTHEPEOPLEOFTHELANDSLEWALLTHEMTH
ATHADCONSPIREDAGAINSTKINGAMONANDTHEPEOPLEOFTHE
LANDMADEJOSIAHHISSONKINGINHISSTEADJOSIAHWASEIG
HTYEARSOLDWHENHEBEGANTOREIGNANDHEREIGNEDTHIRTY
ANDONEYEARSINJERUSALEMANDHEDIDTHATWHICHWASRIGH
TINTHEEYESOFJEHOVAHANDWALKEDINTHEWAYSOFDAVIDHI
SFATHERANDTURNEDNOTASIDETOTHERIGHTHANDORTOTHEL
EFTFORINTHEEIGHTHYEAROFHISREIGNWHILEHEWASYETYO
UNGHEBEGANTOSEEKAFTERTHEGODOFDAVIDHISFATHERAND
INTHETWELFTHYEARHEBEGANTOPURGEJUDAHANDJERUSALE
MFROMTHEHIGHPLACESANDTHEASHERIMANDTHEGRAVENIMA
GESANDTHEMOLTENIMAGESANDTHEYBRAKEDOWNTHEALTARS
OFTHEBAALIMINHISPRESENCEANDTHESUNIMAGESTHATWER
EONHIGHABOVETHEMHEHEWEDDOWNANDTHEASHERIMANDTHE
GRAVENIMAGESANDTHEMOLTENIMAGESHEBRAKEINPIECESA
NDMADEDUSTOFTHEMANDSTREWEDITUPONTHEGRAVESOFTHE
MTHATHADSACRIFICEDUNTOTHEMANDHEBURNTTHEBONESOF
THEPRIESTSUPONTHEIRALTARSANDPURGEDJUDAHANDJERU
SALEMANDSODIDHEINTHECITIESOFMANASSEHANDEPHRAIM
ANDSIMEONEVENUNTONAPHTALIINTHEIRRUINSROUNDABOU
TANDHEBRAKEDOWNTHEALTARSANDBEATTHEASHERIMANDTH
EGRAVENIMAGESINTOPOWDERANDHEWEDDOWNALLTHESUNIM
AGESTHROUGHOUTALLTHELANDOFISRAELANDRETURNEDTOJ
```

Verses in Matrix

The Second Book of the Chronicles 33:24-34:7

That sounds pretty conclusive, doesn't it? I find the passages above meaningful, in that for me, as a numerologist, they contain esoteric messages. Note that the verses are numbered from 33 to 34, and I have elsewhere discussed the numerological significance of these numbers in Diana's life during the last two years of her existence on this earthly plane.

Diana in one of her favourite dresses, the one she dubbed her "Elvis" dress.

Matches for: Diana die Alma Pont head injury note sent Paul

Match: 1 of 1 Skip to Match

ELS (Equidistant Letter Spacing) Matrix

```
DPAULANDCERTAINOTHERPRISONERSTOACENTURIONNAMED
JULIUSOFTHEAUGUSTANBANDANDEMBARKINGINASHIPOFAD
RAMYTTIUMWHICHWASABOUTTOSAILUNTOTHEPLACESONTHE
COASTOFASIAWEPUTTOSEAARISTARCHUSAMACEDONIANOFT
HESSALONICABEINGWITHUSANDTHENEXTDAYWETOUCHEDAT
SIDONANDJULIUSTREATEDPAULKINDLYANDGAVEHIMLEAVE
TOGOUNTOHISFRIENDSANDREFRESHHIMSELFANDPUTTINGT
OSEAFROMTHENCEWESAILEDUNDERTHELEEOFCYPRUSBECAU
SETHEWINDSWERECONTRARYANDWHENWEHADSAILEDACROSS
THESEAWHICHISOFFCILICIAANDPAMPHYLIAWECAMETOMYR
AACITYOFLYCIAANDTHERETHECENTURIONFOUNDASHIPOFA
LEXANDRIASAILINGFORITALYANDHEPUTUSTHEREINANDWH
ENWEHADSAILEDSLOWLYMANYDAYSANDWERECOMEWITHDIFF
ICULTYOVERAGAINSTCNIDUSTHEWINDNOTFURTHERSUFFER
INGUSWESAILEDUNDERTHELEEOFCRETEOVERAGAINSTSALM
ONEANDWITHDIFFICULTYCOASTINGALONGITWECAMEUNTOA
CERTAINPLACECALLEDFAIRHAVENSNIGHWHEREUNTOWASTH
ECITYOFLASEAANDWHENMUCHTIMEWASSPENTANDTHEVOYAG
EWASNOWDANGEROUSBECAUSETHEFASTWASNOWALREADYGON
EBYPAULADMONISHEDTHEMANDSAIDUNTOTHEMSIRSIPERCE
IVETHATTHEVOYAGEWILLBEWITHINJURYANDMUCHLOSSNOT
```

Verses in Matrix

The Acts of the Apostles 27:1-27:10

The matrix above, on the face of it, seems astounding evidence not only for the existence of the note, but also that it was written by Diana. One thing we always have to bear in mind, however, is that if the negation of a statement appears within the same matrix, it will render the matrix and therefore the statement false.

If negating words can be added to the matrix, the statement "Diana die Alma Pont head injury note sent Paul" is necessarily false. Unfortunately, I can only use 10 search words at a time, but if you look at the following matrix, you will see that within those same Biblical passages can be found the word "not".

We can't have it both ways. If The Bible Code is true, it will state all and everything we need to know. We have to look at both the statement of the search and any possible negation within the matrix that might render that statement false. Unless we insert "false", "not", "untrue" and "not true" in conjunction with the matrix, to see if they apply, we cannot hope to reach the right conclusion. The word "not" is completely divorced from the word "note" in the matrix below.

People have pointed out that since Diana's prediction in The Mischon Note (that Charles would marry Tiggy at her demise) bore no fruition; correlatively, her prediction that she would die in a car accident engineered by Charles, is simply an unfortunate coincidence; one that has absolutely no basis in established fact. Her sister Sarah has intimated in her evidence to the court that Diana was relying on her astrologer's predictions rather than any "hard" evidence, in regard to the fear she held for her personal safety.

Matches for: Diana die Alma Pont head injury note not sent Paul

Match: 1 of 1 Skip to Match

ELS (Equidistant Letter Spacing) Matrix

```
DPAULANDCERTAINOTHERPRISONERSTOACENTURIONNAMED
JULIUSOFTHEAUGUSTANBANDANDEMBARKINGINASHIPOFAD
RAMYTTIUMWHICHWASABOUTTOSAILUNTOTHEPLACESONTHE
COASTOFASIAWEPUTTOSEAARISTARCHUSAMACEDONIANOFT
HESSALONICABEINGWITHUSANDTHENEXTDAYWETOUCHEDAT
SIDONANDJULIUSTREATEDPAULKINDLYANDGAVEHIMLEAVE
TOGOUNTOHISFRIENDSANDREFRESHHIMSELFANDPUTTINGT
OSEAFROMTHENCEWESAILEDUNDERTHELEEOFCYPRUSBECAU
SETHEWINDSWERECONTRARYANDWHENWEHADSAILEDACROSS
THESEAWHICHISOFFCILICIAANDPAMPHYLIAWECAMETOMYR
AACITYOFLYCIAANDTHERETHECENTURIONFOUNDASHIPOFA
LEXANDRIASAILINGFORITALYANDHEPUTUSTHEREINANDWH
ENWEHADSAILEDSLOWLYMANYDAYSANDWERECOMEWITHDIFF
ICULTYOVERAGAINSTCNIDUSTHEWINDNOTFURTHERSUFFER
INGUSWESAILEDUNDERTHELEEOFCRETEOVERAGAINSTSALM
ONEANDWITHDIFFICULTYCOASTINGALONGITWECAMEUNTOA
CERTAINPLACECALLEDFAIRHAVENSNIGHWHEREUNTOWASTH
ECITYOFLASEAANDWHENMUCHTIMEWASSPENTANDTHEVOYAG
EWASNOWDANGEROUSBECAUSETHEFASTWASNOWALREADYGON
EBYPAULADMONISHEDTHEMANDSAIDUNTOTHEMSIRSIPERCE
IVETHATTHEVOYAGEWILLBEWITHINJURYANDMUCHLOSSNOT
```

Verses in Matrix

The Acts of the Apostles 27:1-27:10

On the strength of the above, if the Bible Code is true, we are forced to the conclusion that either Paul Burrell has lied, and forged the note, (which seems highly unlikely), or someone else has written the note, pretending to be Diana, in order to deceive Paul.

But what could be the motive? Why would someone want to do such a thing? Perhaps to make her out to be paranoid, of unsound mind? Danger existed in committing such an act, in that it would point the finger of suspicion at a lot of people if something were to happen to her in the wake of the note's delivery, which it did.

Quote. By Gordon Rayner, Chief Reporter 12:01AM GMT 20 Dec 2007 The Telegraph.(England) A handwritten letter in which Diana, Princess of Wales claimed that the Prince of Wales was plotting to kill her so he could marry Tiggy Legge-Bourke, the former nanny to Princes William and Harry, has been shown at the inquest into her death. The note was sent to Paul Burrell, the Princess's then butler, in October 1993, 10 months after her separation from Prince Charles was announced. End of Quote.

Were these the ravings of a mentally ill woman, or the letter a fabrication of a liar, Diana's butler Paul Burrell? Was he bought off by Dodi's father? Or did he truthfully believe the letter he held was from Diana, but was in reality a hoax, designed to convince him of its authenticity? Surely he spoke to Diana in person in the intervening time prior to her death, in order to verify the authenticity of the letter?

Certainly he was prepared to front the inquest and swear on oath that this letter delivered to him was from Diana. If it was written by someone other than Diana, how strange that her death transpired just some months later, in exactly the same circumstances. Could it have been a mere coincidence?

Princess Diana and her butler and close confidante Paul Burrell on a visit to Bosnia on 2nd August, 1997, just prior to her death 29 days later. She was there to publicize her concern for landmines left behind after military conflict, and the subsequent danger they pose to civil life, particularly children. Photograph: Tim Rooke/Rex Features

The Matrix below poses a problem of ambiguity. If we interpret it to mean "Diana die Alma Pont, injury note not sent, Paul lie" that means the note is a fake. If we interpret it to mean "Diana die Alma Pont, injury note not sent Paul, lie" the we get a completely different picture, one which says the note is real and that Diana really did write it and send it to Paul Burrell, her butler, exactly as he claims.

Matches for: Diana die Alma Pont injury note not sent Paul lie

Match: 1 **of 1** Skip to Match

ELS (Equidistant Letter Spacing) Matrix

```
DPAULANDCERTAINOTHERPRISONERSTOACENTURIONNAMED
JULIUSOFTHEAUGUSTANBANDANDEMBARKINGINASHIPOFAD
RAMYTTIUMWHICHWASABOUTTOSAILUNTOTHEPLACESONTHE
COASTOFASIAWEPUTTOSEAARISTARCHUSAMACEDONIANOFT
HESSALONICABEINGWITHUSANDTHENEXTDAYWETOUCHEDAT
SIDONANDJULIUSTREATEDPAULKINDLYANDGAVEHIMLEAVE
TOGOUNTOHISFRIENDSANDREFRESHHIMSELFANDPUTTINGT
OSEAFROMTHENCEWESAILEDUNDERTHELEEOFCYPRUSBECAU
SETHEWINDSWERECONTRARYANDWHENWEHADSAILEDACROSS
THESEAWHICHISOFFCILICIAANDPAMPHYLIAWECAMETOMYR
AACITYOFLYCIAANDTHERETHECENTURIONFOUNDASHIPOFA
LEXANDRIASAILINGFORITALYANDHEPUTUSTHEREINANDWH
ENWEHADSAILEDSLOWLYMANYDAYSANDWERECOMEWITHDIFF
ICULTYOVERAGAINSTCNIDUSTHEWINDNOTFURTHERSUFFER
INGUSWESAILEDUNDERTHELEEOFCRETEOVERAGAINSTSALM
ONEANDWITHDIFFICULTYCOASTINGALONGITWECAMEUNTOA
CERTAINPLACECALLEDFAIRHAVENSNIGHWHEREUNTOWASTH
ECITYOFLASEAANDWHENMUCHTIMEWASSPENTANDTHEVOYAG
EWASNOWDANGEROUSBECAUSETHEFASTWASNOWALREADYGON
EBYPAULADMONISHEDTHEMANDSAIDUNTOTHEMSIRSIPERCE
IVETHATTHEVOYAGEWILLBEWITHINJURYANDMUCHLOSSNOT
```

Verses in Matrix

The Acts of the Apostles 27:1-27:10

Nanny Tiggy Legge Bourke with Charles, William and Harry.

Diana showing her love for the camera, her photogenic face and form, and her acting ability right from childhood. These qualities would one day cause her harassment in her private life. Born into wealth, the daughter of Lord Spencer. Diana would one day exchange this fabulous baby carriage for a magnificent royal one.

Section 8. Diana feared assassination over anti-landmine campaign. Diana 'believed she would be bumped off'. Simone Simmons, the landmines dossier and the alleged call from the grandson of Sir Winston Churchill.

There have been reports that Diana's involvement with the land mine issue caused her to receive death threats; one such threat from no less than a British Member of Parliament. Quote: Diana, Princess of Wales believed she was going to be "bumped off" by MI6 because of her high profile campaign against landmines, the inquest into her death has been told. The biggest and most economically important industry in the UK is 'defence' - the arms industry. The British government, military and intelligence services form an integral part of this industry, and the regime controls and subsidises it, the military-industrial complex.

Her friend Simone Simmons said the Princess was about to "name names" and publish a dossier called “Profiting Out of Misery”. "Top of her list of culprits was the Secret Intelligence Service which she believed was behind the sale of British landmines that were causing so much misery to so many people," Miss Simmons said.

On one occasion the Princess sent her a note which said: "If something happens, MI5 or MI6 will have done it." Diana had been feeling "shaken" since the brakes failed in her car, although fortunately on this occasion nobody was injured. Most drivers never experience brake failure. Diana's luck was unusually bad, especially for somebody with access to the best cars, the best mechanical servicing, and the best drivers that money can buy.

Miss Simmons said the Princess gave her the landmines dossier, which was up to six inches thick, and she kept it for several months under her mattress along with the note. Later, she burned them. She said: "I was more than nervous. If I had the material, I might have been bumped off as well."

At Kensington Palace in February 1997, the Princess asked her to listen in to a phone call she was on in which she was threatened over the landmine campaign, Miss Simmons said. "This person was saying to her that she shouldn't interfere in matters she knew nothing about," she said. The caller told the Princess, "Well, accidents can happen" and she was "very pale" after the conversation. Miss Simmons said the Princess told her the caller was Conservative MP Nicholas Soames. Mr. Soames, the grandson of Winston Churchill, has previously told the inquest that the suggestion was "grotesque and preposterous".

The Honourable Nicholas Soames MP, who emphatically denied the above accusation, Miss Simone Simmons, who was Diana’s spiritual advisor, Insignia of the NSA

Simone testified in court that Diana asked her to listen into a conversation with Nicolas Soames who allegedly made what appeared to be a threatening phone call to Diana over the land mines issue and her campaign against their use by the British Military. Diana charged Simone with the responsibility of keeping the land mines dossier which was up to six inches thick, under the mattress of her bed for safe keeping. In the wake of Diana's untimely death, she said she destroyed it out of fear for her own personal safety.

Matches for: Simone hide Diana note MOD SIS fear die road

Match: 1 **of 1** Skip to Match

ELS (Equidistant Letter Spacing) Matrix

```
DBEYEVALIANTFORSAULYOURLORDISDEADANDALSOTHEHOU
SEOFJUDAHHAVEANOINTEDMEKINGOVERTHEMNOWABNERTHE
SONOFNERCAPTAINOFSAULSHOSTHADTAKENISHBOSHETHTH
ESONOFSAULANDBROUGHTHIMOVERTOMAHANAIMANDHEMADE
HIMKINGOVERGILEADANDOVERTHEASHURITESANDOVERJEZ
REELANDOVEREPHRAIMANDOVERBENJAMINANDOVERALLISR
AELISHBOSHETHSAULSSONWASFORTYYEARSOLDWHENHEBEG
ANTOREIGNOVERISRAELANDHEREIGNEDTWOYEARSBUTTHEH
OUSEOFJUDAHFOLLOWEDDAVIDANDTHETIMETHATDAVIDWAS
KINGINHEBRONOVERTHEHOUSEOFJUDAHWASSEVENYEARSAN
DSIXMONTHSANDABNERTHESONOFNERANDTHESERVANTSOFI
SHBOSHETHTHESONOFSAULWENTOUTFROMMAHANAIMTOGIBE
ONANDJOABTHESONOFZERUIAHANDTHESERVANTSOFDAVIDW
ENTOUTANDMETTHEMBYTHEPOOLOFGIBEONANDTHEYSATDOW
NTHEONEONTHEONESIDEOFTHEPOOLANDTHEOTHERONTHEOT
HERSIDEOFTHEPOOLANDABNERSAIDTOJOABLETTHEYOUNGM
ENIPRAYTHEEARISEANDPLAYBEFOREUSANDJOABSAIDLETT
HEMARISETHENTHEYAROSEANDWENTOVERBYNUMBERTWELVE
FORBENJAMINANDFORISHBOSHETHTHESONOFSAULANDTWEL
VEOFTHESERVANTSOFDAVIDANDTHEYCAUGHTEVERYONEHIS
```

Verses in Matrix

The Second Book of Samuel 2:7-2:15

According to Miss Simmons, the Princess had secret meetings with Tony Blair in the months before he became Prime Minister about becoming a roving ambassador in relation to landmines. "That's what he promised her and she thought he was going to announce it when he was elected. She was very disappointed," Miss Simmons said. End of Quote. "The Insider" mailing list article, 13 January 2008.

Simone Simmons was Diana's spiritual adviser from 1993. If these claims are true, it is very telling indeed that she entrusted vital information to Simmons and Burrell to the exclusion of her two sisters and brother.

The problem with the matrix above, is that "Diana" and "Simone" could be any two people on the globe. I could find no identifying words like "Princess" or "Alma Tunnel/Pont" which would verify their identities within it.

It would seem that there is enough evidence, in consideration of the above matrices, to support Paul Burrell's claim that Diana did indeed write the note, now in the possession of Diana's lawyer, (Lord Mishcon). We could justifiably believe that the note is real and not a fake.

Reference: Ibid. The Telegraph. 26th January, 2012. Judges found there were problems in establishing that Henri Paul, the couple's chauffeur who also died, was drunk at the time of the accident and a mix-up with his post-mortem examination and blood tests created unnecessary delays.

The new French inquiry is being led by Judge Gerard Caddeo, Mr. al Fayed said. Mr. Al Fayed's claims that evidence was withheld centre around a letter - dubbed the Mishcon note - which was written by Diana's divorce lawyer Lord Mishcon after an October 1995 meeting outlining her fears that there was a plot to kill her in a car crash.

In his statement issued Mr. al Fayed said: "His (Judge Caddeo's) interviews of Lord Condon and Mr. Veness are the beginning of a thorough investigation by the French authorities into why Lord Mishcon's letter was not provided to the initial French inquiry.

As the father of Dodi and a good friend of the princess, I trust that the outcome of these French investigations will be that those who are guilty will be properly punished." Lord Mishcon passed on the letter to police in confidence following a meeting with Lord Condon after the Princess's death.

It was only when Diana's former butler, Paul Burrell, produced a note from the princess making very similar allegations in the Daily Mirror in October 2003 that the police agreed to hand the letter over and that December it was sent to the coroner investigating the princess's death. Mr. Al Fayed has always said he believed there had been a cover up. At the UK inquest into the deaths, Mr. Al Fayed implicated many establishment figures in the plot to kill or cover up the murder of the couple. End of quote.

Princess Diana's chef from Kensington Palace, Darren McGrady. He worked for Diana from 1993 until her death in 1997. He now lives in the United States, and has written a book about Diana's favourite recipes.

Prince Harry serving in Afghanistan 2008

Section 9. Hold ups with the ongoing inquiry in France. Royal coroner to visit Diana death scene.

While I am in the process of writing this book, today this article below appeared on the Internet. It has been very difficult to get any word of how the new inquest in France is going. One would think reports of such an important nature would be splashed all over the Internet, but this is not the case at this time.

I guess being a part of the British Commonwealth, loyal subjects to HRH Queen Elizabeth, has its drawbacks at times. And the arrangements for the Queen's Diamond Jubilee are in progress. It strangely looks like the court inquiry is being delayed until that celebration has passed.

Quote: Mail Online Tuesday, Apr 17 2012 12AM

The Royal Coroner holding an inquest into the death of Diana, Princess of Wales, has travelled to Paris where he is expected to visit the scene of the crash which killed her. Michael Burgess will be accompanied by detectives from Scotland Yard as he goes to see for himself the spot in the Alma tunnel.

Mr. Burgess and the police team are also meeting French officers who investigated Diana's death. The brief trip is expected to last only a few days and Scotland Yard Commissioner Sir John Stevens, who Mr. Burgess asked to investigate the circumstances of Diana's death, may visit the French capital at a later date.

Mr. Burgess is expected to visit key places including the Ritz Hotel, from where Diana's car set off, to familiarize himself with the area. It is understood he will probably go into the underpass to see the crash scene. Police are also looking at the 6,000-page report into the tragedy produced by French judge Herve Stephan.

Mr. Burgess opened and adjourned his inquest into Diana's death a month ago - six-and-a-half years after she died on August 31, 1997. The inquest follows persistent claims from Harrods owner Mohamed Fayed, father of Dodi

Fayed who was also killed in the crash, that the couple was murdered by those who wanted to stop their relationship.

As well as being Coroner of the Royal Household Mr. Burgess is Coroner for Surrey and is therefore also holding an inquest into Dodi Fayed's death. Diana, Dodi and driver Henri Paul were all killed when the Mercedes limousine they were in crashed.
End of Quote.

Reference:http://www.dailymail.co.uk/news/article-207475/Royal-coroner-visit-Diana-death-scene.html#ixzz1sFglP17c

The note Paul Burrell allegedly produced, seemed, (if not to implicate Charles in her actual death), to implicate him in scheming to arrange a car accident designed to, at the very least, seriously injure her.

Lord Mischon the incorruptible lawyer, who died in 2006. Perhaps one reason for the success of the solicitor Lord Mishcon, who has died aged 90, was his sense that the law should be there for everybody. He had an old-fashioned sense of the responsibilities of being a lawyer in the days before law practices became commercial exercises. Masterful with private clients, he was a man who would extricate them from whatever problem they had dug for themselves. He did not preach to them and he had an ability to set an agenda.

A Labour member for Lambeth of the Greater London council (GLC) from 1964 to 1967 - and chairman of its general purposes committee - Victor Mishcon was created a life peer in 1978 during James Callaghan's administration. He was effectively shadow chancellor (1983-92) when he was opposition spokesman on legal affairs in the Lords.

Later, unlike many lawyer MPs and peers, he was not in fear of New Labour's Lord Irvine. His position served the Law Society well, and Mishcon acted as an intermediary during the 1999 debates on the access to justice bill, which was seen as severely restricting the availability of legal aid, of which Mishcon was a great supporter. As a member of the Lords, he chose his subjects carefully, speaking only on topics he knew and understood. As a result, everyone listened. He would have been pleased to be regarded as old Labour.

Armoiries république française

If we are to believe the following matrix, the outcome of this court case may be no different from the one before, or is the finding below only relating to the first inquest that came before the court? After all, the reason the case has been brought before the court again, is because certain information was allegedly concealed, in that it was never revealed. A crime of deliberate omission is a serious one.

Matches for: France judge find Diana road death plan not true

Match: 1 of 1 Skip to Match

ELS (Equidistant Letter Spacing) Matrix

```
NEMIESFROMBEFORETHEEANDIWILLMAKETHEEANAMELIKEU
NTOTHENAMEOFTHEGREATONESTHATAREINTHEEARTHANDIW
ILLAPPOINTAPLACEFORMYPEOPLEISRAELANDWILLPLANTT
HEMTHATTHEYMAYDWELLINTHEIROWNPLACEANDBEMOVEDNO
MORENEITHERSHALLTHECHILDRENOFWICKEDNESSWASTETH
EMANYMOREASATTHEFIRSTANDASFROMTHEDAYTHATICOMMA
NDEDJUDGESTOBEOVERMYPEOPLEISRAELANDIWILLSUBDUE
ALLTHINEENEMIESMOREOVERITELLTHEETHATJEHOVAHWIL
LBUILDTHEEAHOUSEANDITSHALLCOMETOPASSWHENTHYDAY
SAREFULFILLEDTHATTHOUMUSTGOTOBEWITHTHYFATHERST
HATIWILLSETUPTHYSEEDAFTERTHEEWHOSHALLBEOFTHYSO
NSANDIWILLESTABLISHHISKINGDOMHESHALLBUILDMEAHO
USEANDIWILLESTABLISHHISTHRONEFOREVERIWILLBEHIS
FATHERANDHESHALLBEMYSONANDIWILLNOTTAKEMYLOVING
KINDNESSAWAYFROMHIMASITOOKITFROMHIMTHATWASBEFO
RETHEEBUTIWILLSETTLEHIMINMYHOUSEANDINMYKINGDOM
FOREVERANDHISTHRONESHALLBEESTABLISHEDFOREVERAC
CORDINGTOALLTHESEWORDSANDACCORDINGTOALLTHISVIS
IONSODIDNATHANSPEAKUNTODAVIDTHENDAVIDTHEKINGWE
NTINANDSATBEFOREJEHOVAHANDHESAIDWHOAMIOJEHOVAH
GODANDWHATISMYHOUSETHATTHOUHASTBROUGHTMETHUSFA
RANDTHISWASASMALLTHINGINTHINEEYESOGODBUTTHOUHA
```

Verses in Matrix

The First Book of the Chronicles 17:8-17:7

From the same verses in the Bible, the following matrix emerges, "France find Diana road death plan true time two", but of course, to this one also can be applied the word "not" which appears in these verses, negating it entirely. But if the word "not" only applies to the first inquest, (and it seems logical that it would) then the French court will find that there actually was a plan of conspiracy to kill Diana and Dodi, or at the very least, injure them. The plan may have gone awry, and the fatal outcome could have been unintended. Still, anyone considering such criminal action surely must have considered all possible outcomes?

The Beautiful Arc de Triomphe, on the banks of the Seine River, Paris. Such a romantically revered city and river; but now global icons many will always connect with the deaths of Princess Diana and Dodi Fayed, star-crossed lovers, unacceptable to the world in which they lived.

Matches for: France find Diana road death plan true time two

Match: 1 of 1 Skip to Match

ELS (Equidistant Letter Spacing) Matrix

THATAREINTHEEARTHANDIWILLAPPOINTAPLACEFORMYPEO
PLEISRAELANDWILLPLANTTHEMTHATTHEYMAYDWELLINTHE
IROWNPLACEANDBEMOVEDNOMORENEITHERSHALLTHECHILD
RENOFWICKEDNESSWASTETHEMANYMOREASATTHEFIRSTAND
ASFROMTHEDAYTHATICOMMANDEDJUDGESTOBEOVERMYPEOP
LEISRAELANDIWILLSUBDUEALLTHINEENEMIESMOREOVERI
TELLTHEETHATJEHOVAHWILLBUILDTHEEAHOUSEANDITSHA
LLCOMETOPASSWHENTHYDAYSAREFULFILLEDTHATTHOUMUS
TGOTOBEWITHTHYFATHERSTHATIWILLSETUPTHYSEEDAFTE
RTHEEWHOSHALLBEOFTHYSONSANDIWILLESTABLISHHISKI
NGDOMHESHALLBUILDMEAHOUSEANDIWILLESTABLISHHIST
HRONEFOREVERIWILLBEHISFATHERANDHESHALLBEMYSONA
NDIWILLNOTTAKEMYLOVINGKINDNESSAWAYFROMHIMASITO
OKITFROMHIMTHATWASBEFORETHEEBUTIWILLSETTLEHIMI
NMYHOUSEANDINMYKINGDOMFOREVERANDHISTHRONESHALL
BEESTABLISHEDFOREVERACCORDINGTOALLTHESEWORDSAN
DACCORDINGTOALLTHISVISIONSODIDNATHANSPEAKUNTOD
AVIDTHENDAVIDTHEKINGWENTINANDSATBEFOREJEHOVAHA
NDHESAIDWHOAMIOJEHOVAHGODANDWHATISMYHOUSETHATT
HOUHASTBROUGHTMETHUSFARANDTHISWASASMALLTHINGIN
THINEEYESOGODBUTTHOUHASTSPOKENOFTHYSERVANTSHOU

Verses in Matrix

The First Book of the Chronicles 17:8-17:17

Ritz Hotel Paris, from which Diana and Dodi travelled to the Alma Pont in the Mercedes. The Imperial suite in the Ritz Hotel, owned by Mohamed Al Fayed, where Diana and Dodi ate their last meal together.

Dodi with his Dad Mohamed, who raised him as a single father after his divorce. Dodi was born on 15th April, 1955, in Alexandria, Egypt, when his Dad was just 21 years old.

Section 10. The Duke of Edinburgh and Prince Charles – Mohamed Fayed's claims of their involvement in Diana's death, by inducing action from the Secret Intelligence Service (SIS) and the Ministry of Defence (MOD).

What of the astounding claim by Mohamed Al Fayed, that the Duke of Edinburgh had, through this expressed attitude to Diana in the wake of her divorce from Charles, influenced SIS (MI6) to take action against her, i.e.: to dispose of her?

What of the charge levelled against Charles in the Mischon note? There would have had to be a massive shift in the Duke's feelings for her. We have plenty of written evidence that shows the pair "got on very well together", had a "fond and affectionate relationship", yet there have been claims the Duke "wrote cruel and derogatory letters" to her in the years prior to her death.

Less than 5 minutes after this heart-wrenching security (CCTV) camera footage was taken, Dodi was dead and Diana lay dying in the back of the Fayed black Mercedes in the Alma Tunnel.

The Duke of Edinburgh with the radiantly beautiful mother-to-be Princess Diana, prior to the birth of her first child, son and heir to the throne, Prince William. He reportedly told Diana when she lost HRH title after her divorce, "Don't worry Mummy. When I am King I will give it back to you." And I don't doubt that he will.

There has been plenty of evidence that, even though relations between Diana and Charles were strained initially, their friendship seemed, to the world at large, to be quite intact following the divorce. That is why the revelations in the letter to Paul Burrell and "The Mischon Note", held by Lord Mischon, Diana's lawyer, which included her instructions and confidences to him prior to her death, so astounded the world.

Above is the photograph taken by Lord Snowdon, Princess Margaret's photographer husband and the SIS BUILDING, commonly known as the MI6, located at 85 Albert Embankment in the south western part of central London, on the bank of the River Thames beside Vauxhall Bridge. MI5 is formally known as the Secret Service, and deals with British internal affairs, while MI6 is known as the Secret Intelligence Service (SIS) and deals with external or foreign affairs. MI5 is the British security service, while MI6 is the British foreign intelligence service.

MI6 are British spies. "MI5/MI6" were the original designations when both organisations came under the War Office, now the MOD - "MI" stands for military intelligence. Their official names (acquired in the 30's) are the Security Service (MI5) and SIS, the Secret Intelligence Service (MI6). The former is responsible to the Home Office and the latter to the Foreign Office.

Horrifyingly archaic as the prospect of Diana being murdered by the SIS, MOD, and especially by a member of the royal family sounds, there is no doubt such things have happened in our not too distant past. If we want to take a stroll down history lane, we encounter Elizabeth I's mother, Anne Boleyn, who was executed in 1536 by her husband Henry VIII and his Cromwell court. She was his second wife, and he went on to execute his fifth wife, Anne's first cousin, Catherine Howard, just six years later, in 1542. "King Henry the Eighth, to six wives he was wedded. One died, one survived, two divorced, two beheaded."

Elizabeth I in turn imprisoned in The Tower her (supposed) much loved Sir Walter Raleigh and his new wife Elizabeth Throckmorton. Raleigh's crime of treason? He married without his monarch's permission. He had the good sense to realize the parlous position he was in, and wrote the Queen adoring poems from his cell in The Tower of London. Raleigh had once famously laid down his cloak for her over a mud puddle, so her royal slippers would not spoil.

He even had the good grace to bring back the potato and tobacco to England from The Americas in the 16th Century. All this did not save him from the noose. He was hanged after Elizabeth's death in 1618. No doubt his track record as a criminal with "prior" (incurring the Queen's displeasure) did the poor fellow no favours at the trial which led to his demise.

We are now in the 21st Century, yet in the evolutionary scheme of things, not light years ahead of those times. Consequently, many people around the world wondered, "Could it have happened again, even in this modern age?" I decided to run some searches.

Henry VIII with his children, one of whom is the red-haired Elizabeth I.

When using long search words, I have found it is necessary to cut them into separate pieces, comprised of groups of syllables. This invariably, for some researchers, weakens the results of the matrix. In the above results, "mount" in the ordinary Biblical text intersects with "batten". Philip Mountbatten is the Duke of Edinburgh's commonly known name. For some, like Rips and Drosnin, this would indicate added strength to the matrix results.

Significance of the number 16 in Diana's demise.

The number 16 has long been associated with royalty, in combination with both accidents befalling them and treason alleged against them. It can of course, be subjects committing treason against them, beneath the disguise of "assassidents". Thus, this number alone will not tell us whether or not Diana and Dodi met with an accident or a plot to murder Diana. Look carefully at the Tarot Symbolism for the number 16. (XVI=Roman numerals for the number 16)

Note the crown falling from the heads of the couple. Note the tower, meant to symbolize the Tower of London, where Queens like Anne Boleyn, after being tried and convicted, awaited the executioner. Usually the charge was a trumped up one, to get rid of an inconvenient Queen. Alleged infidelity to her husband the King on the part of the Queen, or even incest historically was charged. Sometimes the charges happened to be true, sometimes not.

What happened to Charles in 19**79**? - 7+9=16/7. See below. What happened to Charles in 19**88**? Master 88 is the highest vibration of the number 16/7: (8+8=16/7). He managed to survive an avalanche that killed his friend. What happened to Charles' ex-wife in 19**97**? (9+7=16/7)

Prince in fatal ski incident. Unhurt Charles joins in rescue attempt after close friend is buried by avalanche on Swiss mountainside. The Guardian, Friday 11 March 1988 11.07 GMT

Quote: A distraught Prince of Wales took part yesterday in a desperate attempt to rescue one of his close friends, who was killed after being buried by an avalanche which swept down a mountain in the Swiss resort of Klosters where the royal party was skiing. Prince Charles, who was not hurt in the incident, joined other members of the party in a frantic bid to dig out Major Hugh Lindsay, who was caught in the cascade of tons of freshly fallen snow which shot down Gotschnagrat Mountain in the fashionable ski resort.

A second member of the party, who was also trapped in the avalanche, Mrs Charles Palmer-Tomkinson, sustained leg injuries. Major Lindsay, a former equerry to the Queen, was found to be dead on arrival at Davos hospital, to where he and Mrs Palmer-Tomkinson were helicoptered by the Red Cross after the accident. The incident occurred at 2.45 pm yesterday when the Prince of Wales, Major Lindsay, Mr and Mrs Palmer-Tomkinson, a Swiss guide and a Swiss police officer were in a stationary position off-piste near the Wang run.

According to Buckingham Palace, the avalanche started above them and all of the royal party were able to take avoiding action except the two victims, who were caught in the sudden slide of snow which had been building up on the mountain yesterday during a fresh and heavy fall. Eye-witnesses said that Prince Charles, who was lifted off the slope by a second helicopter, was visibly distressed. The second helicopter pilot was quoted by locally based reporters as saying the Prince was weeping. Neither the Princess of Wales nor the Duchess of York, who are also on the skiing holiday, were on the mountain when the avalanche occurred. They were both in the royal party's chalet in Davos.

The slopes on Gotschnagrat Mountain are amongst the steepest in Switzerland and are rarely open to the general public. They are regarded as suitable only for experienced skiers like Prince Charles. Before yesterday's tragedy there had been a heavy snowfall on the mountain, although it was not thought last night that the sudden change in weather conditions had led the Klosters authorities to issue an avalanche warning.

Major Lindsay was 34. His wife, Sarah, who is expecting a baby, works in the press office at Buckingham Palace. A career army officer, he returned to his full-time army duties in 1986 after his spell in service to the Queen as an equerry. Mrs Lindsay was being comforted by friends in London last night. At her family's home in Godalming, Surrey, the housekeeper said: 'Sarah's baby is due in May and this must be a terrible shock for her. They were only married two years ago.'

The Queen and the Duke of Edinburgh were informed about the incident yesterday afternoon during an official engagement in south London. Last night Buckingham Palace said they did not know whether the royal party in Klosters would cut short their visit, due to last until March 20, but it was expected they would return home within the next 48 hours.

Avalanches can be set off without warning by temperature increases, new falls of snow, or skiers and climbers. Powder snow avalanches follow falls of fine snow and can reach speeds of 120 mph. They often contain as much air as snow and their greatest threat is death by suffocation. The Swiss government has a federal avalanche institute at Davos, which warns when strong winds and snowfalls increase the avalanche risk. On Sunday, skiing conditions in the Davos area were reported to be 'superb.'

Prince Charles, who has been skiing since 1963, had an accident near the Wang run in 1979 when he lost control on the slopes and landed heavily although without injury on his left hip. Since 1980 he and all members of the Royal Family have worn electric bleepers in case they are buried in snow drifts. Since their marriage he and Princess Diana have been skiing at least once a year. End of Quote.

Prince Charles Not Harry's Real Father; Ex Diana Lover Keeps Silent Because of Death Threats From Royal Family

Posted on May 31st, 2009 by HisHighness in IUC:Entertainment, IUC:Exclusive

Quote: Prince Harry is the toast of NYC this week. Throngs of folks have lined the streets of Manhattan to greet Britain's popular prince with open arms. Little do they realize that Harry could soon be stripped of his royal title because he's not the biological son of Prince Charles, an IUC Investigation has revealed.

A long time employee of Harry's mother Princess Diana told IUC that the Royal Family was involved in a massive cover-up to hide the fact that Diana's ex lover James Hewitt is Harry's real father. According to the source Prince Philip threatened Hewitt's life if he didn't go along with the cover-up. "They made him lie about the timeline," the source told IUC. "Prince Philip told Hewitt he would destroy him if it ever leaked out. It's impossible that Charles is Harry's real father. Hewitt was on the scene as Diana's lover two years before Harry was born. Diana stopped having sex with Charles years before Harry was born. Harry looks exactly like Hewitt."

The massive cover-up involved Hewitt lying to the world about when his dalliance actually began with Diana. Originally he told the world he met Diana in 1986. Harry was born in 1984. Under hypnosis for a TV interview Hewitt admitted he met Diana in 1981 or 1982, had sexual relations with her then – some two years before Harry was born. A relative of Hewitt told IUC that privately Hewitt has always believed he's Harry's dad but has denied it in public because he fears for his life. He also wants to protect his son from being dethroned.

"Bloody hell, Harry should call James his real father," the relative said. "It's the biggest lie in Britain since Neville Chamberlain assured the world some seven decades ago that Britain and Germany would never go to war. Prince Philip told James he'd kill him if he ever admitted the truth. That's why he keeps denying it." The relative added that this is another example of why the Royal Family should be abolished. "They're the biggest crooks and liars in the world," he said. "All the evidence clearly demonstrates that James is Harry's real father. Just the way they killed Diana they'll kill James. You'll see, one day his body will be found mysteriously in a hotel room and the Royals will try to convince the world that James committed suicide." End of Quote.

I could find no matches for “James” or “Hewitt” in combination with the name of Prince Harry or Prince Henry. The best I could get was the title “Captain” in syllables. I could find no reference to Charles or William and the throne. However, there were references to Harry and the throne. Diana has gone on record as saying William told her he does not want to be King. But little children say a lot of things they later recant as adults. Mohamed Al Fayed believes William thinks there was a conspiracy to kill his mother, that he will one day just walk away from the throne, that William is too free spirited to stick with the staid life British royalty will require of him, and instead will head for the freedom of America. Well, he’s dyed in the wool British, groomed for royalty and kingship, so we’ll have to wait and see, won’t we?

Matches for: Prince Harry son Cap tain lover Mum Princess not true

Match: 1 **of 1** Skip to Match

ELS (Equidistant Letter Spacing) Matrix

```
NLIFTEDUPTHEHEADOFJEHOIACHINKINGOFJUDAHANDBROU
GHTHIMFORTHOUTOFPRISONANDHESPAKEKINDLYTOHIMAND
SETHISTHRONEABOVETHETHRONEOFTHEKINGSTHATWEREWI
THHIMINBABYLONANDCHANGEDHISPRISONGARMENTSANDJE
HOIACHINDIDEATBREADBEFOREHIMCONTINUALLYALLTHED
AYSOFHISLIFEANDFORHISALLOWANCETHEREWASACONTINU
ALALLOWANCEGIVENHIMBYTHEKINGOFBABYLONEVERYDAYA
PORTIONUNTILTHEDAYOFHISDEATHALLTHEDAYSOFHISLIF
ETHELAMENTATIONSOFJEREMIAHHOWDOTHTHECITYSITSOL
ITARYTHATWASFULLOFPEOPLE!SHEISBECOMEASAWIDOWTH
ATWASGREATAMONGTHENATIONS!SHETHATWASAPRINCESSA
MONGTHEPROVINCESISBECOMETRIBUTARY!SHEWEEPETHSO
REINTHENIGHTANDHERTEARSAREONHERCHEEKSAMONGALLH
ERLOVERSSHEHATHNONETOCOMFORTHERALLHERFRIENDSHA
VEDEALTTREACHEROUSLYWITHHERTHEYAREBECOMEHERENE
MIESJUDAHISGONEINTOCAPTIVITYBECAUSEOFAFFLICTIO
NANDBECAUSEOFGREATSERVITUDESHEDWELLETHAMONGTHE
NATIONSSHEFINDETHNORESTALLHERPERSECUTORSOVERTO
OKHERWITHINTHESTRAITSTHEWAYSOFZIONDOMOURNBECAU
SENONECOMETOTHESOLEMNASSEMBLYALLHERGATESAREDES
```

The Book of the Prophet Jeremiah 52:31-
The Lamentations of Jeremiah 1:4

Matches for: Prince Henry Diana son King one day not true false

Match: 1 **of 1** Skip to Match

ELS (Equidistant Letter Spacing) Matrix

```
NDITCAMETOPASSTHATWHENTHEARMYOFTHECHALDEANSWAS
BROKENUPFROMJERUSALEMFORFEAROFPHARAOHSARMYTHEN
JEREMIAHWENTFORTHOUTOFJERUSALEMTOGOINTOTHELAND
OFBENJAMINTORECEIVEHISPORTIONTHEREINTHEMIDSTOF
THEPEOPLEANDWHENHEWASINTHEGATEOFBENJAMINACAPTA
INOFTHEWARDWASTHEREWHOSENAMEWASIRIJAHTHESONOFS
HELEMIAHTHESONOFHANANIAHANDHELAIDHOLDONJEREMIA
HTHEPROPHETSAYINGTHOUARTFALLINGAWAYTOTHECHALDE
ANSTHENSAIDJEREMIAHITISFALSEIAMNOTFALLINGAWAYT
OTHECHALDEANSBUTHEHEARKENEDNOTTOHIMSOIRIJAHLAI
DHOLDONJEREMIAHANDBROUGHTHIMTOTHEPRINCESANDTHE
PRINCESWEREWROTHWITHJEREMIAHANDSMOTEHIMANDPUTH
IMINPRISONINTHEHOUSEOFJONATHANTHESCRIBEFORTHEY
HADMADETHATTHEPRISONWHENJEREMIAHWASCOMEINTOTHE
DUNGEONHOUSEANDINTOTHECELLSANDJEREMIAHHADREMAI
NEDTHEREMANYDAYSTHENZEDEKIAHTHEKINGSENTANDFETC
HEDHIMANDTHEKINGASKEDHIMSECRETLYINHISHOUSEANDS
AIDISTHEREANYWORDFROMJEHOVAHANDJEREMIAHSAIDTHE
REISHESAIDALSOTHOUSHALTBEDELIVEREDINTOTHEHANDO
FTHEKINGOFBABYLONMOREOVERJEREMIAHSAIDUNTOKINGZ
EDEKIAHWHEREINHAVEISINNEDAGAINSTTHEEORAGAINSTT
HYSERVANTSORAGAINSTTHISPEOPLETHATYEHAVEPUTMEIN
```

The Book of the Prophet Jeremiah 37:11 - 37:18

The double negative in the matrix above renders the statement “Prince Henry Diana son King one day” necessarily true. The word “lovers” also appears in the top one and the linear text makes particularly interesting reading.

Entrance to the Ritz Hotel

Section 11. Was Diana Jemima Khan's half-sister?

"Diana intended to start a campaign in a Palestinian refugee camp in Gaza or the West Bank," Nicholas Davies (National Journal)

Lady Diana's mother was Frances Shand-Kydd, who was born Frances Ruth Roche. Frances had Jewish origins. (Davidic Ancestry of Prince William.) Lady Diana's father may have been the Jewish billionaire James Goldsmith. Mostaque Ali drew our attention to this possibility. On 7 September 2007, the Independent (Ireland) produced an article entitled: A Murky Question, by Declan Lynch (The Mail also covered the story. Were Diana and Jemima sisters? Mail Online). (www.dailymail.co.uk/.)

According to Declan Lynch, Princess Diana's father was Sir James Goldsmith. Sir James was the son of Frank Goldsmith, the descendant of a Jewish banking family from Frankfurt. He multiplied his fortune in the United States during the 1980s (Billionaire with a Cause vanityfair.com)

"He inherited many of his traits from his father Frank Goldsmith, the descendant of a distinguished Jewish banking family from Frankfurt once as famous as the Rothschilds." (Obituary: Sir James Goldsmith - People, News - The Independent)

The "Goldsmith" version "had been circulating in aristocratic circles for a long time." Officially, Diana was the daughter of the Earl Spencer and Frances Shand-Kydd. Sources have long maintained that Goldsmith was conducting an affair with Frances around the time that Diana was conceived.

Nobody denies that the affair took place, "at a time when Frances was deeply unhappy in her marriage to the Earl, who was 'drinking heavily' and 'being beastly towards her'". Tina Brown, author of The Diana Chronicle, suggests it was a long-running affair. In Brown's version, there is strong support for the idea of Goldsmith being the father of Lady Diana, though she can't prove it.

Zak Goldsmith, son of Sir James Goldsmith. Diana has a physical resemblance to Sir James' children: Zak, Jemima Khan (nee Goldsmith), Diana's best friend, born January 1974, was more than 13 years her junior, and Ben Goldsmith.

Benjamin Goldsmith above, (born in 1980) married Kate Emma Rothschild (born in 1982), daughter of the late Amschel Rothschild.

Section 12. The possibility of Diana's father being not the 8th Earl (Edward) John Spencer, but Jewish billionaire Sir James Goldsmith.

It has always been known that Frances, Diana's mother had Jewish origins, and therefore, Diana had known Jewish heritage. Yet, some might have thought it strange that Diana's greatest friend, (from many accounts), was a young lady nearly thirteen years her junior: Jemima Khan, formerly Goldsmith.

Since it is evidently common knowledge in high circles, that Diana's mother had an affair with Jemima's father James Goldsmith at around the time Diana was conceived, let us look at The Bible Code to see if there is any justification for the rumours that Jemima and Diana were in fact half-sisters.

Zak Goldsmith left, Princess Diana aged 19, right.

Another fact that gives weight to the idea that Diana was conceived as a result of an extra marital affair, is that when Frances left her husband, reportedly she took only Diana with her, yet Charles was nearly three years younger than his sister and the couple had five children in all.

A bitter legal battle ensued, with the then Lord Spencer gaining custody of Diana. In order to differentiate one man from another, the religion and the name are the obvious choices for key search words. Lord Spencer was Anglican, Sir James Goldsmith was Jewish. Perhaps the most telling matrices are those revealed from the search words below.

Prince William in Royal Navy uniform 2008.

Matches for: Lady Diana Mum hide Jew father daughter Diana

Match: 1 of 2 Skip to Match

ELS (Equidistant Letter Spacing) Matrix

```
HIMMICHALHISDAUGHTERTOWIFEANDSAULSAWANDKNEWTHA
TJEHOVAHWASWITHDAVIDANDMICHALSAULSDAUGHTERLOVE
DHIMANDSAULWASYETTHEMOREAFRAIDOFDAVIDANDSAULWA
SDAVIDSENEMYCONTINUALLYTHENTHEPRINCESOFTHEPHIL
ISTINESWENTFORTHANDITCAMETOPASSASOFTENASTHEYWE
NTFORTHTHATDAVIDBEHAVEDHIMSELFMOREWISELYTHANAL
LTHESERVANTSOFSAULSOTHATHISNAMEWASMUCHSETBYAND
SAULSPAKETOJONATHANHISSONANDTOALLHISSERVANTSTH
ATTHEYSHOULDSLAYDAVIDBUTJONATHANSAULSSONDELIGH
TEDMUCHINDAVIDANDJONATHANTOLDDAVIDSAYINGSAULMY
FATHERSEEKETHTOSLAYTHEENOWTHEREFOREIPRAYTHEETA
KEHEEDTOTHYSELFINTHEMORNINGANDABIDEINASECRETPL
ACEANDHIDETHYSELFANDIWILLGOOUTANDSTANDBESIDEMY
FATHERINTHEFIELDWHERETHOUARTANDIWILLCOMMUNEWIT
HMYFATHEROFTHEEANDIFISEEAUGHTIWILLTELLTHEEANDJ
ONATHANSPAKEGOODOFDAVIDUNTOSAULHISFATHERANDSAI
DUNTOHIMLETNOTTHEKINGSINAGAINSTHISSERVANTAGAIN
STDAVIDBECAUSEHEHATHNOTSINNEDAGAINSTTHEEANDBEC
AUSEHISWORKSHAVEBEENTOTHEEWARDVERYGOODFORHEPUT
HISLIFEINHISHANDANDSMOTETHEPHILISTINEANDJEHOVA
```

Verses in Matrix

The First Book of Samuel 18:27-19:2

Le Caprice restaurant in London Diana frequented. She always sat at Table 7, the number of rest peace and quiet, ruled by the sign of Cancer.

Matches for: Diana die Alma Pont Dad Jew Goldsmith

Match: 1 of 1 Skip to Match

ELS (Equidistant Letter Spacing) Matrix

```
DBYHIMASLESSTHANNOTHINGANDVANITYTOWHOMTHENWILL
YELIKENGODORWHATLIKENESSWILLYECOMPAREUNTOHIMTH
EIMAGEAWORKMANHATHCASTITANDTHEGOLDSMITHOVERLAY
ETHITWITHGOLDANDCASTETHFORITSILVERCHAINSHETHAT
ISTOOIMPOVERISHEDFORSUCHANOBLATIONCHOOSETHATRE
ETHATWILLNOTROTHESEEKETHUNTOHIMASKILFULWORKMAN
TOSETUPAGRAVENIMAGETHATSHALLNOTBEMOVEDHAVEYENO
TKNOWNHAVEYETNOTHEARDHATHITNOTBEENTOLDYOUFROMT
HEBEGINNINGHAVEYENOTUNDERSTOODFROMTHEFOUNDATIO
NSOFTHEEARTHITISHETHATSITTETHABOVETHECIRCLEOFT
HEEARTHANDTHEINHABITANTSTHEREOFAREASGRASSHOPPE
RSTHATSTRETCHETHOUTTHEHEAVENSASACURTAINANDSPRE
ADETHTHEMOUTASATENTTODWELLINTHATBRINGETHPRINCE
STONOTHINGTHATMAKETHTHEJUDGESOFTHEEARTHASVANIT
YYEATHEYHAVENOTBEENPLANTEDYEATHEYHAVENOTBEENSO
WNYEATHEIRSTOCKHATHNOTTAKENROOTINTHEEARTHMOREO
VERHEBLOWETHUPONTHEMANDTHEYWITHERANDTHEWHIRLWI
NDTAKETHTHEMAWAYASSTUBBLETOWHOMTHENWILLYELIKEN
METHATISHOULDBEEQUALTOHIMSAITHTHEHOLYONELIFTUP
YOUREYESONHIGHANDSEEWHOHATHCREATEDTHESETHATBRI
```

Verses in Matrix

The Book of the Prophet Isaiah 40:17-40:25

La Famiglia, Italian restaurant Diana frequented in Chelsea, London, where her picture hangs over her usual table.

Something has displeased them

Matches for: Lady Diana Mum keep secret Jew father Diana

Match: 1 of 1 Skip to Match

ELS (Equidistant Letter Spacing) Matrix

```
MICHALSAULSDAUGHTERLOVEDHIMANDSAULWASYETTHEMOR
EAFRAIDOFDAVIDANDSAULWASDAVIDSENEMYCONTINUALLY
THENTHEPRINCESOFTHEPHILISTINESWENTFORTHANDITCA
METOPASSASOFTENASTHEYWENTFORTHTHATDAVIDBEHAVED
HIMSELFMOREWISELYTHANALLTHESERVANTSOFSAULSOTHA
THISNAMEWASMUCHSETBYANDSAULSPAKETOJONATHANHISS
ONANDTOALLHISSERVANTSTHATTHEYSHOULDSLAYDAVIDBU
TJONATHANSAULSSONDELIGHTEDMUCHINDAVIDANDJONATH
ANTOLDDAVIDSAYINGSAULMYFATHERSEEKETHTOSLAYTHEE
NOWTHEREFOREIPRAYTHEETAKEHEEDTOTHYSELFINTHEMOR
NINGANDABIDEINASECRETPLACEANDHIDETHYSELFANDIWI
LLGOOUTANDSTANDBESIDEMYFATHERINTHEFIELDWHERETH
OUARTANDIWILLCOMMUNEWITHMYFATHEROFTHEEANDIFISE
EAUGHTIWILLTELLTHEEANDJONATHANSPAKEGOODOFDAVID
UNTOSAULHISFATHERANDSAIDUNTOHIMLETNOTTHEKINGSI
NAGAINSTHISSERVANTAGAINSTDAVIDBECAUSEHEHATHNOT
SINNEDAGAINSTTHEEANDBECAUSEHISWORKSHAVEBEENTOT
HEEWARDVERYGOODFORHEPUTHISLIFEINHISHANDANDSMOT
ETHEPHILISTINEANDJEHOVAHWROUGHTAGREATVICTORYFO
RALLISRAELTHOUSAWESTITANDDIDSTREJOICEWHEREFORE
```

Verses in Matrix

The First Book of Samuel 18:28-19:5

There were also matches for "Lady Diana Mum keep secret Diana son heir King."

Dodi on the set of Chariots of Fire

Matches for: Lady Diana Mum hide Jew birth two sons Princes

Match: 1 of 1 Skip to Match

ELS (Equidistant Letter Spacing) Matrix

```
THEMIGHTBETHEKINGSSONINLAWANDSAULGAVEHIMMICHAL
HISDAUGHTERTOWIFEANDSAULSAWANDKNEWTHATJEHOVAHW
ASWITHDAVIDANDMICHALSAULSDAUGHTERLOVEDHIMANDSA
ULWASYETTHEMOREAFRAIDOFDAVIDANDSAULWASDAVIDSEN
EMYCONTINUALLYTHENTHEPRINCESOFTHEPHILISTINESWE
NTFORTHANDITCAMETOPASSASOFTENASTHEYWENTFORTHTH
ATDAVIDBEHAVEDHIMSELFMOREWISELYTHANALLTHESERVA
NTSOFSAULSOTHATHISNAMEWASMUCHSETBYANDSAULSPAKE
TOJONATHANHISSONANDTOALLHISSERVANTSTHATTHEYSHO
ULDSLAYDAVIDBUTJONATHANSAULSSONDELIGHTEDMUCHIN
DAVIDANDJONATHANTOLDDAVIDSAYINGSAULMYFATHERSEE
KETHTOSLAYTHEENOWTHEREFOREIPRAYTHEETAKEHEEDTOT
HYSELFINTHEMORNINGANDABIDEINASECRETPLACEANDHID
ETHYSELFANDIWILLGOOUTANDSTANDBESIDEMYFATHERINT
HEFIELDWHERETHOUARTANDIWILLCOMMUNEWITHMYFATHER
OFTHEEANDIFISEEAUGHTIWILLTELLTHEEANDJONATHANSP
AKEGOODOFDAVIDUNTOSAULHISFATHERANDSAIDUNTOHIML
ETNOTTHEKINGSINAGAINSTHISSERVANTAGAINSTDAVIDBE
CAUSEHEHATHNOTSINNEDAGAINSTTHEEANDBECAUSEHISWO
RKSHAVEBEENTOTHEEWARDVERYGOODFORHEPUTHISLIFEIN
```

Verses in Matrix

The First Book of Samuel 18:27-19:2

Section 13. The Spencer family's estrangement from their sister, their unwillingness to believe her when she told them her life was in danger, and their seeming willingness to write her off as paranoid, delusional, manipulative and deceitful.

Now, after her demise, and in the face of mounting evidence, they continue to assert the latter.

In consideration of the claim by Diana's brother Charles Spencer (and others) that, (leaving aside their acknowledgment of the concern Diana had for her personal safety), she was mentally ill before she died, and had been for some time, we need to take that into consideration, and look at things from that angle. Her death, occurring chapter and verse according to her forewarning, is written off as an unfortunate coincidence.

There seems to be ample evidence to show that Diana had been mentally ill for a large part of her life, given her battles with anorexia nervosa and bulimia which were public knowledge for years before her death. A troubled childhood no doubt caused or largely contributed to this condition, which in all likelihood worsened with the stress of being in the public arena as future Queen of England.

Yet these mental conditions by themselves are unlikely to cause compulsive lying, a manipulative and deceitful personality, and extreme paranoia - her belief that she was about to be injured or murdered by her ex-husband, MOD and SIS.

Even if we are to believe that Diana's condition had not improved, that in fact her mental health had declined (as seems to have been suggested by her brother Charles), at the time of her death, this does not necessarily preclude the belief that she was murdered. In fact, it would make it that much easier for anyone to dispose of her, knowing full well any warnings of her impending murder she gave to those around her would not be believed, perhaps not even in a court of law, should it come to that.

The Royal Courts of Justice London.

Section 14. Diana's childhood and her prior affinity with the royal family.

Diana grew up in Park House, practically next door to The Queen. Park House is right next to the Royals' Sandringham estate. The Spencer children were therefore no strangers to the Princes Charles, Andrew, Edward and their sister Anne. Prince Andrew, who was the closest in age to Diana, was her childhood playmate.

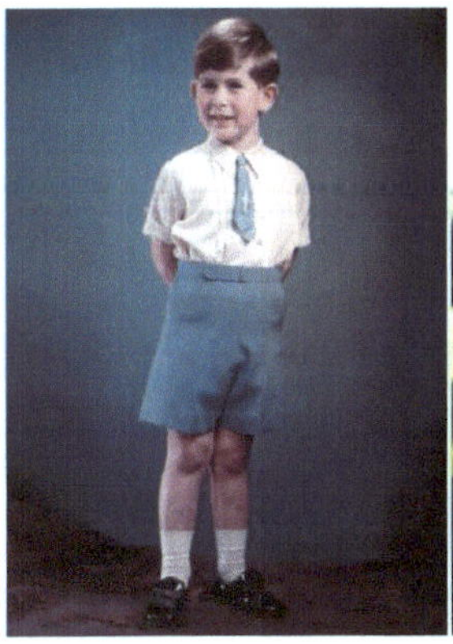

Princes Charles and Andrew, riding the tractor. Diana with her brother Charles.

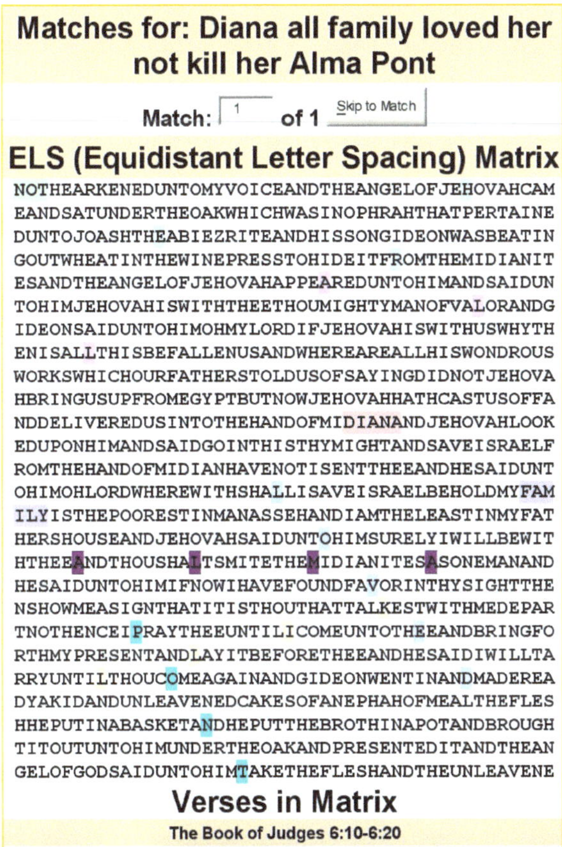

Matches for: Diana all family loved her not kill her Alma Pont

Match: 1 of 1 Skip to Match

ELS (Equidistant Letter Spacing) Matrix

```
NOTHEARKENEDUNTOMYVOICEANDTHEANGELOFJEHOVAHCAM
EANDSATUNDERTHEOAKWHICHWASINOPHRAHTHATPERTAINE
DUNTOJOASHTHEABIEZRITEANDHISSONGIDEONWASBEATIN
GOUTWHEATINTHEWINEPRESSTOHIDEITFROMTHEMIDIANIT
ESANDTHEANGELOFJEHOVAHAPPEAREDUNTOHIMANDSAIDUN
TOHIMJEHOVAHISWITHTHEETHOUMIGHTYMANOFVALORANDG
IDEONSAIDUNTOHIMOHMYLORDIFJEHOVAHISWITHUSWHYTH
ENISALLTHISBEFALLENUSANDWHEREAREALLHISWONDROUS
WORKSWHICHOURFATHERSTOLDUSOFSAYINGDIDNOTJEHOVA
HBRINGUSUPFROMEGYPTBUTNOWJEHOVAHHATHCASTUSOFFA
NDDELIVEREDUSINTOTHEHANDOFMIDIANANDJEHOVAHLOOK
EDUPONHIMANDSAIDGOINTHISTHYMIGHTANDSAVEISRAELF
ROMTHEHANDOFMIDIANHAVENOTISENTTHEEANDHESAIDUNT
OHIMOHLORDWHEREWITHSHALLISAVEISRAELBEHOLDMYFAM
ILYISTHEPOORESTINMANASSEHANDIAMTHELEASTINMYFAT
HERSHOUSEANDJEHOVAHSAIDUNTOHIMSURELYIWILLBEWIT
HTHEEANDTHOUSHALTSMITETHEMIDIANITESASONEMANAND
HESAIDUNTOHIMIFNOWIHAVEFOUNDFAVORINTHYSIGHTTHE
NSHOWMEASIGNTHATITISTHOUTHATTALKESTWITHMEDEPAR
TNOTHENCEIPRAYTHEEUNTILICOMEUNTOTHEEANDBRINGFO
RTHMYPRESENTANDLAYITBEFORETHEEANDHESAIDIWILLTA
RRYUNTILTHOUCOMEAGAINANDGIDEONWENTINANDMADEREA
DYAKIDANDUNLEAVENEDCAKESOFANEPHAHOFMEALTHEFLES
HHEPUTINABASKETANDHEPUTTHEBROTHINAPOTANDBROUGH
TITOUTUNTOHIMUNDERTHEOAKANDPRESENTEDITANDTHEAN
GELOFGODSAIDUNTOHIMTAKETHEFLESHANDTHEUNLEAVENE
```

Verses in Matrix

The Book of Judges 6:10-6:20

The love shared is apparent: Duke of Edinburgh, Princes William and Harry. Prince Charles and Princess Diana. She had just turned 20, he was 32, when they married on 29th July, 1981.

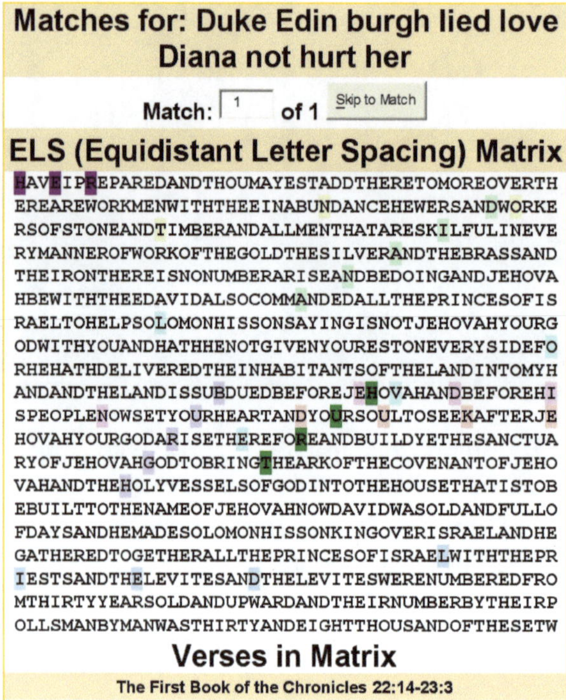

Matches for: Duke Edin burgh lied love Diana not hurt her

Match: 1 of 1 Skip to Match

ELS (Equidistant Letter Spacing) Matrix

HAVEIPREPAREDANDTHOUMAYESTADDTHERETOMOREOVERTH
EREAREWORKMENWITHTHEEINABUNDANCEHEWERSANDWORKE
RSOFSTONEANDTIMBERANDALLMENTHATARESKILFULINEVE
RYMANNEROFWORKOFTHEGOLDTHESILVERANDTHEBRASSAND
THEIRONTHEREISNONUMBERARISEANDBEDOINGANDJEHOVA
HBEWITHTHEEDAVIDALSOCOMMANDEDALLTHEPRINCESOFIS
RAELTOHELPSOLOMONHISSONSAYINGISNOTJEHOVAHYOURG
ODWITHYOUANDHATHHENOTGIVENYOURESTONEVERYSIDEFO
RHEHATHDELIVEREDTHEINHABITANTSOFTHELANDINTOMYH
ANDANDTHELANDISSUBDUEDBEFOREJEHOVAHANDBEFOREHI
SPEOPLENOWSETYOURHEARTANDYOURSOULTOSEEKAFTERJE
HOVAHYOURGODARISETHEREFOREANDBUILDYETHESANCTUA
RYOFJEHOVAHGODTOBRINGTHEARKOFTHECOVENANTOFJEHO
VAHANDTHEHOLYVESSELSOFGODINTOTHEHOUSETHATISTOB
EBUILTTOTHENAMEOFJEHOVAHNOWDAVIDWASOLDANDFULLO
FDAYSANDHEMADESOLOMONHISSONKINGOVERISRAELANDHE
GATHEREDTOGETHERALLTHEPRINCESOFISRAELWITHTHEPR
IESTSANDTHELEVITESANDTHELEVITESWERENUMBEREDFRO
MTHIRTYYEARSOLDANDUPWARDANDTHEIRNUMBERBYTHEIRP
OLLSMANBYMANWASTHIRTYANDEIGHTTHOUSANDOFTHESETW

Verses in Matrix

The First Book of the Chronicles 22:14-23:3

In the matrix above, "Edin" crosses "burgh" and "Duke", giving added strength to the matrix. The words "she was ill" also come up in the same section as above, but it is not possible to run a search (using the computer system I had) with more than ten keywords input at the one time. This of necessity restricts the results.

Of course, even if the Duke hated Diana, that does not mean he killed her by engineering her death. (Yet the matrix above could be interpreted to imply that he also lied about not harming her.) In looking at the matrix below, the same is absolutely true of Charles. Even if he came to dislike Diana intensely after they parted, does not mean he would ever dream of harming her, the mother of his young sons.

Diana was embalmed and buried on an island in the middle of the lake at Althorp. In her casket, she wears a black dress and holds a rosary in her hands. The memorial is close to the tomb where she lies buried.

Matches for: Husband lied she was ill kill Mum Diana Alma Pont

Match: 1 of 4 Skip to Match

ELS (Equidistant Letter Spacing) Matrix

```
SORWHOISHETHATGAVETHEETHISAUTHORITYANDHEANSWER
EDANDSAIDUNTOTHEMIALSOWILLASKYOUAQUESTIONANDTE
LLMETHEBAPTISMOFJOHNWASITFROMHEAVENORFROMMENAN
DTHEYREASONEDWITHTHEMSELVESSAYINGIFWESHALLSAYF
ROMHEAVENHEWILLSAYWHYDIDYENOTBELIEVEHIMBUTIFWE
SHALLSAYFROMMENALLTHEPEOPLEWILLSTONEUSFORTHEYA
REPERSUADEDTHATJOHNWASAPROPHETANDTHEYANSWEREDT
HATTHEYKNEWNOTWHENCEITWASANDJESUSSAIDUNTOTHEMN
EITHERTELLIYOUBYWHATAUTHORITYIDOTHESETHINGSAND
HEBEGANTOSPEAKUNTOTHEPEOPLETHISPARABLEAMANPLAN
TEDAVINEYARDANDLETITOUTTOHUSBANDMENANDWENTINTO
ANOTHERCOUNTRYFORALONGTIMEANDATTHESEASONHESENT
UNTOTHEHUSBANDMENASERVANTTHATTHEYSHOULDGIVEHIM
OFTHEFRUITOFTHEVINEYARDBUTTHEHUSBANDMENBEATHIM
ANDSENTHIMAWAYEMPTYANDHESENTYETANOTHERSERVANTA
NDHIMALSOTHEYBEATANDHANDLEDHIMSHAMEFULLYANDSEN
THIMAWAYEMPTYANDHESENTYETATHIRDANDHIMALSOTHEYW
OUNDEDANDCASTHIMFORTHANDTHELORDOFTHEVINEYARDSA
IDWHATSHALLIDOIWILLSENDMYBELOVEDSONITMAYBETHEY
WILLREVERENCEHIMBUTWHENTHEHUSBANDMENSAWHIMTHEY
REASONEDONEWITHANOTHERSAYINGTHISISTHEHEIRLETUS
KILLHIMTHATTHEINHERITANCEMAYBEOURSANDTHEYCASTH
```

Verses in Matrix

The Gospel According to St. Luke 20:2-20:14

In the matrix above, the words "husband", "was" and "kill" are in the ordinary Biblical text, perhaps weakening their import. Neither "husband" nor "kill" intersect or cross any of the text of the matrix, lessening their value perhaps. The words "lied" and "Diana" intersect, perhaps giving weight to their value.

In the matrix above, the words "not family" are in the ordinary text, which might render the matrix impotent in the view of some. However, the word "family" intersects with "loved" in the matrix, which many researchers would consider significant, enhancing its value.

Section 15. Prince Michael of Kent and his alleged involvement with the SIS and Diana's death.

Some people, in their desire to lay the blame at the feet of royalty for Diana's death, have gone so far as to say Prince Michael of Kent and his Masonic connections within the SIS are responsible for the Princess's demise. Although this sounds like a ridiculous assertion, let's leave no stone unturned in our search for revealing matrices.

The title "Prince" and the name "Michael" appear in the ordinary Biblical text, for some lessening their power; while the word of the matrix that spells "Diana" intersects with "Michael".

Although there is nothing here to separate this Diana with other Diana's by placing her in the Alma Tunnel at the time of her death, the full title of Prince Michael of Kent comes up. The double negative renders the assertion "it is not true Prince Michael Kent SIS kill Diana" a false one. It also renders true the assertion "it is not false that Prince Michael Kent SIS kill Diana."

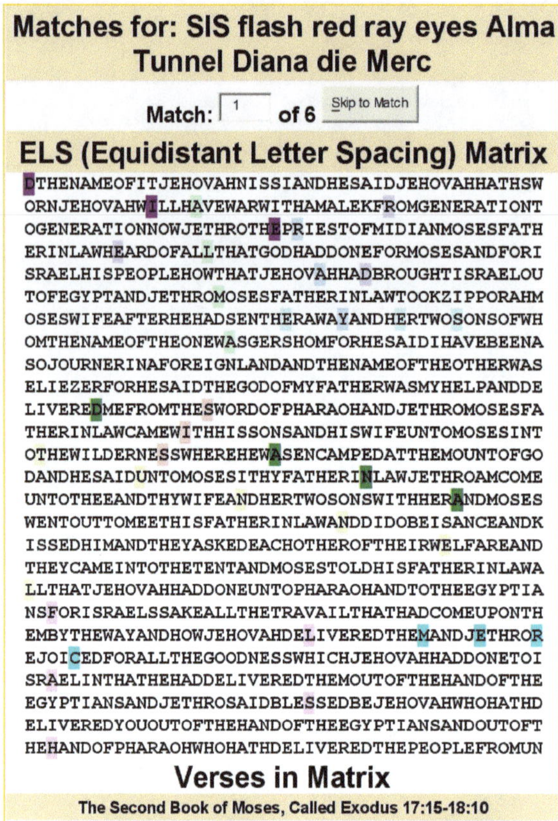

Prince Michael of Kent, right.

William shows Dad around the base

Matches for: Prince Michael Kent SIS kill Diana true not false

Match: 1 of 1 Skip to Match

ELS (Equidistant Letter Spacing) Matrix

```
NHESHALLSTRETCHFORTHHISHANDALSOUPONTHECOUNTRIE
SANDTHELANDOFEGYPTSHALLNOTESCAPEBUTHESHALLHAVE
POWEROVERTHETREASURESOFGOLDANDOFSILVERANDOVERA
LLTHEPRECIOUSTHINGSOFEGYPTANDTHELIBYANSANDTHEE
THIOPIANSSHALLBEATHISSTEPSBUTTIDINGSOUTOFTHEEA
STANDOUTOFTHENORTHSHALLTROUBLEHIMANDHESHALLGOF
ORTHWITHGREATFURYTODESTROYANDUTTERLYTOSWEEPAWA
YMANYANDHESHALLPLANTTHETENTSOFHISPALACEBETWEEN
THESEAANDTHEGLORIOUSHOLYMOUNTAINYETHESHALLCOME
TOHISENDANDNONESHALLHELPHIMANDATTHATTIMESHALLM
ICHAELSTANDUPTHEGREATPRINCEWHOSTANDETHFORTHECH
ILDRENOFTHYPEOPLEANDTHERESHALLBEATIMEOFTROUBLE
SUCHASNEVERWASSINCETHEREWASANATIONEVENTOTHATSA
METIMEANDATTHATTIMETHYPEOPLESHALLBEDELIVEREDEV
ERYONETHATSHALLBEFOUNDWRITTENINTHEBOOKANDMANYO
FTHEMTHATSLEEPINTHEDUSTOFTHEEARTHSHALLAWAKESOM
ETOEVERLASTINGLIFEANDSOMETOSHAMEANDEVERLASTING
CONTEMPTANDTHEYTHATAREWISESHALLSHINEASTHEBRIGH
TNESSOFTHEFIRMAMENTANDTHEYTHATTURNMANYTORIGHTE
OUSNESSASTHESTARSFOREVERANDEVERBUTTHOUODANIELS
```

Verses in Matrix

The Book of Daniel 11:41-12:3

Prince Michael of Kent is a Grand Master in the order of Freemasonry. Freemasons, like Jews, have been among those groups discriminated against, and have had to cope with persecution from far right extremists like the Nazis.

While the number is not accurately known, it is estimated that between 80,000 and 200,000 Freemasons were killed under the Nazi regime. Masonic concentration camp inmates were graded as political prisoners and were forced to wear an inverted red triangle.

Reference: http://en.wikipedia.org/wiki/Freemasonry.

It therefore defies belief that such a man would want to harm a female relative, a mother of young children, especially someone who may have been the victim of persecution for her beliefs, religious and/or political. Yet if The Bible Code is true, it astoundingly yields the results above.

Rumours continue to circulate regarding Prince Michael of Kent's influence over members of the SIS and on Royal affairs.

Princess Michael of Kent's 'Toyboy' is silenced.

May 21st, 2012
Author: Richard Henley Davis

Quote: Princess Michael of Kent's very close friend, Mikhail Kravchenko, has been murdered in a contract killing in Moscow. 46 year old Kravchenko's body was found in the Moscow suburb of Peredelkino after being shot five times to the torso then dragged from his Mercedes Benz to be shot in the head.

Police are not ruling out the murder as a contract killing relating to a business deal or dispute and believe it was the doing of two professional gunmen. The Russian millionaire and Princess Michael of Kent have, in the past, been linked as lovers. But both parties denied that their relationship was anything other than a close platonic friendship.

However in 2006 speculation ran rife over the nature of their close friendship after photographs emerged of Princess Michael and Kravchenko holding hands and taking gondola rides together in Venice, a trip which saw the Princess allegedly book into a very high class hotel under an alias.

Conspiracy websites are warming up to this news seeing as there are some rather startling similarities to the death of Princess Diana's lover, Dodi Fayed and that of Mikhail Kravchenko. i.e. both men were romantically linked to royals to the detriment of the reputation of the British royal family and both died in an incident involving a Mercedes Benz.

For a long time conspiracy researchers have questioned the role played by the royal family in the death of Lady Diana and the lack of interrogation of the royal family after her death. Diana famously wrote a note to a close friend saying that she feared for her life and that she believed her husband, the Prince of Wales or his family, may try to have her killed in a car crash yet the Royal family faced no such police scrutiny.

The internet is awash with conspiracy theories relating to Diana's death some of which implicate Prince Michael through his influence as one of the highest ranking (if not the most high ranking) Freemasons in the world. Many senior ranking police officers are openly known to be Freemasons and so the potential influence of Prince Michael on police investigations has been a point raised by conspiracy researchers many times in the past. But all this speculation will mask a tragic death because when you put aside all the speculation over who and why, you are left with the tragic death of a man, haunted by the deaths of his wife and family in a car crash in 2001.

Reference: http://www.economicvoice.com/princess-michael-of-kents-toyboy-is-silenced/50029753#ixzz217dxoqO0

Section 16. Andanson, the white Fiat Uno, a mysterious and horrific death.

Andanson (above) was a journalist, purportedly in the Alma Tunnel on the night Diana died, driving a white Fiat Uno. Some scientific tests returned positive matches with the paint on Diana and Dodi's car, and a white Fiat Uno. Andanson was found in May 2000, deep in a forest in France. He was in the driver's seat of his burnt out car, no keys were in the locked vehicle. He was decapitated, reportedly from the force of the blaze, with his head, which had a bullet hole in it, lying beside him, between the front seats. Charges were laid at the feet of Jean-Paul Andanson: that he was a member of the UKN, a covert section of MI6, the SIS, that is comprised of journalists who feed information to the SIS, so they can keep tabs on the activities of certain Very Important Persons.

Daisies suited her, as did roses. The word "daisy" carries a Master 22 vibration, Diana's Master Number.

Diana also had a numerological connection with the rose. The word "rose" carries a 21/3 vibration, the number appearing in Diana's life chart, ruling her life from birth to age 27. Known as The English Rose, she was especially fond of long-stemmed, yellow roses.

ROSE	**ROSES**	**DAISY**
9615	**9 61 51**	**41917**
= 21/3	**= Master 22/4**	**= Master 22/4.**

In light of the following evidence, the findings of the English court could be seen to be lacking, in that they attributed no blame at all to the alleged white Fiat Uno in the crash. For the court to have verified existence of the alleged vehicle in the tunnel that fatal night, might have entailed and therefore admitted at least one of SIS/MI6/MOD/MI5, French intel, USA CIA was involved.

The report continued: "The comparative analysis between the infrared spectra characterizing the black polymer taken from the vehicle's fender, and the trace taken from the door of the Mercedes, show that their absorption bands are identical."

In short, despite the French investigators' endorsement of Andanson's alibi, the forensic tests strongly suggested that his car may have been the white Fiat Uno involved in the fatal crash.

Reference: http://www.rense.com/general2/dian.htm

King Henry and Anne Boleyn

Section 17. John Macnamara, the Harrods director of security, and a retired senior Scotland Yard supervisor of investigations.

"Mr. Andanson had for some time been a prime suspect who had relentlessly pursued Diana and Dodi prior to their arrival in Paris. We have always believed that Andanson was at the scene and that more investigation should have been done into his possible involvement. We believe that his death is no coincidence and that this is a line of inquiry which may help to discover the truth. Was Mr. Andanson killed because of what he knew? This is a question we want answered."

If either Andanson or others were responsible for flashing a blinding light into the face of the driver and others in the vehicle, I thought it might come up in a matrix. There were no matches for "laser" or "infra" but the words "beam" "flash" "red" and "ray" came up, along with the all important placing of them in the Alma Tunnel on the night Diana died.

Below is a simple infrared LED flash light device capable of emitting a very bright light, seen over a mile away. These are used by police forces, NASA shuttle crews, Secret Service personnel, as well as ordinary people all around the world, every day. Even something as simple as this device could cause trouble on the highway in the wrong hands, but if we are to believe Tomlinson, the SIS devices are much more sophisticated and powerful.

A strobe light fitted into a headlamp. Conceivably, one could also be fitted into a tail light, the operation of which could blind a driver travelling behind the vehicle. Witnesses reported seeing "a very bright light" and hearing "a loud bang" before the crash.

White Tower of London, Traitors Gate, Tower Bridge.

Buckingham Palace, British Houses of Parliament

Matches for: SIS flash red ray eyes Alma Tunnel when Diana died

Match: 1 of 7 Skip to Match

ELS (Equidistant Letter Spacing) Matrix

```
RHEAVENANDMOSESBUILTANALTARANDCALLEDTHENAMEOFI
TJEHOVAHNISSIANDHESAIDJEHOVAHHATHSWORNJEHOVAHW
ILLHAVEWARWITHAMALEKFROMGENERATIONTOGENERATION
NOWJETHROTHEPRIESTOFMIDIANMOSESFATHERINLAWHEAR
DOFALLTHATGODHADDONEFORMOSESANDFORISRAELHISPEO
PLEHOWTHATJEHOVAHHADBROUGHTISRAELOUTOFEGYPTAND
JETHROMOSESFATHERINLAWTOOKZIPPORAHMOSESWIFEAFT
ERHEHADSENTHERAWAYANDHERTWOSONSOFWHOMTHENAMEOF
THEONEWASGERSHOMFORHESAIDIHAVEBEENASOJOURNERIN
AFOREIGNLANDANDTHENAMEOFTHEOTHERWASELIEZERFORH
ESAIDTHEGODOFMYFATHERWASMYHELPANDDELIVEREDMEFR
OMTHESWORDOFPHARAOHANDJETHROMOSESFATHERINLAWCA
MEWITHHISSONSANDHISWIFEUNTOMOSESINTOTHEWILDERN
ESSWHEREHEWASENCAMPEDATTHEMOUNTOFGODANDHESAIDU
NTOMOSESITHYFATHERINLAWJETHROAMCOMEUNTOTHEEAND
THYWIFEANDHERTWOSONSWITHHERANDMOSESWENTOUTTOME
ETHISFATHERINLAWANDDIDOBEISANCEANDKISSEDHIMAND
THEYASKEDEACHOTHEROFTHEIRWELFAREANDTHEYCAMEINT
OTHETENTANDMOSESTOLDHISFATHERINLAWALLTHATJEHOV
AHHADDONEUNTOPHARAOHANDTOTHEEGYPTIANSFORISRAEL
SSAKEALLTHETRAVAILTHATHADCOMEUPONTHEMBYTHEWAYA
NDHOWJEHOVAHDELIVEREDTHEMANDJETHROREJOICEDFORA
LLTHEGOODNESSWHICHJEHOVAHHADDONETOISRAELINTHAT
HEHADDELIVEREDTHEMOUTOFTHEHANDOFTHEEGYPTIANSAN
DJETHROSAIDBLESSEDBEJEHOVAHWHOHATHDELIVEREDYOU
OUTOFTHEHANDOFTHEEGYPTIANSANDOUTOFTHEHANDOFPHA
```

Verses in Matrix

The Second Book of Moses, Called Exodus 17:14-18:10

Prince Charles with second wife, Camilla, The Duchess of Cornwall.

Section 18. Cluster bombs and the Diana campaign.

BAE (London) has in the past been criticized for its role in the production of cluster bombs, due to the long term risk of serious injury and death they cause to civilians (they behave similarly to land mines). However, after pressure campaigns from various human rights groups, BAE recently stated it no longer produces land mines or cluster bombs. However, at the time of Diana's death, these cluster bombs and land mines would still have been in production in the UK.

Reference: http://en.wikipedia.org/wiki/BAE_Systems

Rank	Country	Defence Spending ($ Billion)	World Share (%)	% of GDP, as at 2010
1	United States	698.0	42.8	4.8
2	China	119.0	7.3	2.1
3	United Kingdom	59.6	3.7	2.3

2001-12 Rank Arms Exporter	Supplier	2010 -2012 $US Millions
1	United States	8641
2	Russia	6039
3	Germany	2340
4	France	1834
5	United Kingdom	1054

Britain is the world's fifth largest arms exporter. Any disruption to the flow of arms would cost the country dearly. The bottom table above indicates values expressed in millions of US dollars. As at 2010 - 2012, Britain spent $US 1,054 million. Picture: cluster bombs.

The information is from the Stockholm International Peace Research Institute.
Reference: http://en.wikipedia.org/wiki/Arms_industry

The defence of a country is, quite naturally, a big ticket item. The BAE Flying Mantis UAS: (Unmanned Air System) flying as of 2013. A French and British united government effort, it cost £40 million, but nets a fortune for those who produce armaments for war.

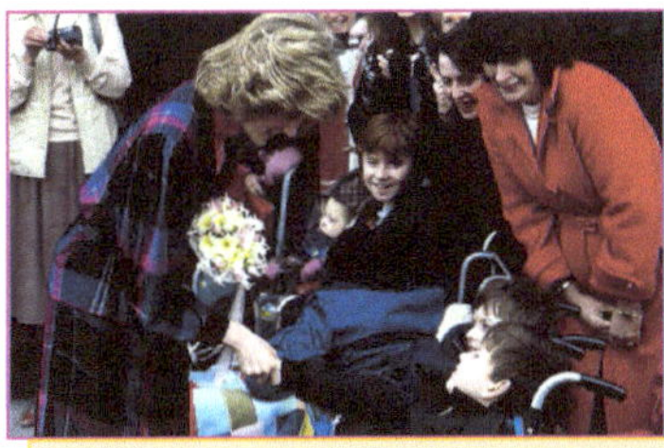

Matches for: Diana Fayed die SIS MOD secret plot Fiat car tip

Match: 1 of 1 Skip to Match

ELS (Equidistant Letter Spacing) Matrix

```
DVERSARIESIAMBECOMEAREPROACHYEAUNTOMYNEIGHBORS
EXCEEDINGLYANDAFEARTOMINEACQUAINTANCETHEYTHATD
IDSEEMEWITHOUTFLEDFROMMEIAMFORGOTTENASADEADMAN
OUTOFMINDIAMLIKEABROKENVESSELFORIHAVEHEARDTHED
EFAMINGOFMANYTERRORONEVERYSIDEWHILETHEYTOOKCOU
NSELTOGETHERAGAINSTMETHEYDEVISEDTOTAKEAWAYMYLI
FEBUTITRUSTEDINTHEEOJEHOVAHISAIDTHOUARTMYGODMY
TIMESAREINTHYHANDDELIVERMEFROMTHEHANDOFMINEENE
MIESANDFROMTHEMTHATPERSECUTEMEMAKETHYFACETOSHI
NEUPONTHYSERVANTSAVEMEINTHYLOVINGKINDNESSLETME
NOTBEPUTTOSHAMEOJEHOVAHFORIHAVECALLEDUPONTHEEL
ETTHEWICKEDBEPUTTOSHAMELETTHEMBESILENTINSHEOLL
ETTHELYINGLIPSBEDUMBWHICHSPEAKAGAINSTTHERIGHTE
OUSINSOLENTLYWITHPRIDEANDCONTEMPTOHHOWGREATIST
HYGOODNESSWHICHTHOUHASTLAIDUPFORTHEMTHATFEARTH
EEWHICHTHOUHASTWROUGHTFORTHEMTHATTAKEREFUGEINT
HEEBEFORETHESONSOFMEN!INTHECOVERTOFTHYPRESENCE
WILTTHOUHIDETHEMFROMTHEPLOTTINGSOFMANTHOUWILTK
EEPTHEMSECRETLYINAPAVILIONFROMTHESTRIFEOFTONGU
ESBLESSEDBEJEHOVAHFORHEHATHSHOWEDMEHISMARVELLO
```

Verses in Matrix

The Book of Psalms Book I Psalm 1 31:11-31:20

Section 19. Mohamed Al Fayed a victim of racism and xenophobia - or is there more here than meets the eye?

Mohamed Al Fayed was born in Alexandria, which ancient city housed the famed library, founded 2,342 years ago in 331 BC; was the birthplace of a disparate group of VIP's: Ptolemy, Euclid, Diophantus, and more recently, Nasser and Omar Sharif.

In an intriguing aside, Fayed married into a family that produced armaments for war. Dodi's mother was Samira Kashoggi, sister of Turkish Saudi-born arms dealer billionaire Adnan Khashoggi. Could this familial connection have worried the British establishment, particularly the SIS? Could it have concerned the CIA, and French Intel?

Would they have feared Fayed political influence over Diana, particularly if she married into the family and produced children? Was Mohamed Al Fayed and consequently Dodi judged guilty by association: Mohamed reportedly Quote: is a former business partner of Cairo attorney El-Amir Atta, father of accused 9/11 suicide hijacker Mohamed el-Amir Awad el-Sayed Atta, who drove and crashed a Boeing 767 into the North Tower of the World Trade Center.

Reference: http://aangirfan.blogspot.com.au/2007/12/al-fayed-and-mohamed-atta.html
Is Tomlinson's claim true, that they even feared religious influence over Diana – that the prospect of her possibly becoming a Muslim was abhorrent to British keepers of the throne, more than alarmed them?

Other newspaper reports claim that Princess Diana and Dodi Fayed were under surveillance by MI6, the CIA, the USA's National Security Agency, the NSA, Mossad and France's DGSE. The NSA (National Security Agency) has over 1050 transcripts covering eavesdropped intercepts of Princess Diana's phone calls.

Mohamed Al Fayed, father of Dodi, left. Photo of Dodi by CNN, right. Note Tower Bridge in the background.

Diana shortly before she died, raising awareness, campaigning against the use of landmines by the British military and others, which undoubtedly aroused political ire in some, admiration in others.

Matches for: Diana Fayed die secret SIS plot not true all lies

Match: 1 of 1 Skip to Match

ELS (Equidistant Letter Spacing) Matrix

```
DVERSARIESIAMBECOMEAREPROACHYEAUNTOMYNEIGHBORS
EXCEEDINGLYANDAFEARTOMINEACQUAINTANCETHEYTHATD
IDSEEMEWITHOUTFLEDFROMMEIAMFORGOTTENASADEADMAN
OUTOFMINDIAMLIKEABROKENVESSELFORIHAVEHEARDTHED
EFAMINGOFMANYTERRORONEVERYSIDEWHILETHEYTOOKCOU
NSELTOGETHERAGAINSTMETHEYDEVISEDTOTAKEAWAYMYLI
FEBUTITRUSTEDINTHEEOJEHOVAHISAIDTHOUARTMYGODMY
TIMESAREINTHYHANDDELIVERMEFROMTHEHANDOFMINEENE
MIESANDFROMTHEMTHATPERSECUTEMEMAKETHYFACETOSHI
NEUPONTHYSERVANTSAVEMEINTHYLOVINGKINDNESSLETME
NOTBEPUTTOSHAMEOJEHOVAHFORIHAVECALLEDUPONTHEEL
ETTHEWICKEDBEPUTTOSHAMELETTHEMBESILENTINSHEOLL
ETTHELYINGLIPSBEDUMBWHICHSPEAKAGAINSTTHERIGHTE
OUSINSOLENTLYWITHPRIDEANDCONTEMPTOHHOWGREATIST
HYGOODNESSWHICHTHOUHASTLAIDUPFORTHEMTHATFEARTH
EEWHICHTHOUHASTWROUGHTFORTHEMTHATTAKEREFUGEINT
HEEBEFORETHESONSOFMEN!INTHECOVERTOFTHYPRESENCE
WILTTHOUHIDETHEMFROMTHEPLOTTINGSOFMANTHOUWILTK
EEPTHEMSECRETLYINAPAVILIONFROMTHESTRIFEOFTONGU
ESBLESSEDBEJEHOVAHFORHEHATHSHOWEDMEHISMARVELLO
```

Verses in Matrix

The Book of Psalms Book I Psalm 1 31:11-31:20

Diana in protective gear at the entrance to a mine field of unexploded cluster bombs, a threat to life and limb negligently left, the aftermath of war.

Matches for: Diana Fayed die secret MOD plot not true all lies

Match: 1 **of 1** Skip to Match

ELS (Equidistant Letter Spacing) Matrix

```
DVERSARIESIAMBECOMEAREPROACHYEAUNTOMYNEIGHBORS
EXCEEDINGLYANDAFEARTOMINEACQUAINTANCETHEYTHATD
IDSEEMEWITHOUTFLEDFROMMEIAMFORGOTTENASADEADMAN
OUTOFMINDIAMLIKEABROKENVESSELFORIHAVEHEARDTHED
EFAMINGOFMANYTERRORONEVERYSIDEWHILETHEYTOOKCOU
NSELTOGETHERAGAINSTMETHEYDEVISEDTOTAKEAWAYMYLI
FEBUTITRUSTEDINTHEEOJEHOVAHISAIDTHOUARTMYGODMY
TIMESAREINTHYHANDDELIVERMEFROMTHEHANDOFMINEENE
MIESANDFROMTHEMTHATPERSECUTEMEMAKETHYFACETOSHI
NEUPONTHYSERVANTSAVEMEINTHYLOVINGKINDNESSLETME
NOTBEPUTTOSHAMEOJEHOVAHFORIHAVECALLEDUPONTHEEL
ETTHEWICKEDBEPUTTOSHAMELETTHEMBESILENTINSHEOLL
ETTHELYINGLIPSBEDUMBWHICHSPEAKAGAINSTTHERIGHTE
OUSINSOLENTLYWITHPRIDEANDCONTEMPTOHHOWGREATIST
HYGOODNESSWHICHTHOUHASTLAIDUPFORTHEMTHATFEARTH
EEWHICHTHOUHASTWROUGHTFORTHEMTHATTAKEREFUGEINT
HEEBEFORETHESONSOFMEN!INTHECOVERTOFTHYPRESENCE
WILTTHOUHIDETHEMFROMTHEPLOTTINGSOFMANTHOUWILTK
EEPTHEMSECRETLYINAPAVILIONFROMTHESTRIFEOFTONGU
ESBLESSEDBEJEHOVAHFORHEHATHSHOWEDMEHISMARVELLO
```

Verses in Matrix

The Book of Psalms Book I Psalm 1 31:11-31:20

Now, just hold your horses a minute there. N.B. If the above words can be found within the last two matrices, then of course, so can these words: “Diana Fayed die secret MOD SIS plot, all true, not lies.” You will have to make up your own mind how you are going to interpret these findings. Perhaps the way you will interpret them will be on an overall assessment that includes the findings of the other matrices in this book.

Section 20. The year before Princess Diana died, she was alienated from her brother Earl Spencer.

The speech given by Earl Charles Spencer at Diana’s funeral impressed everyone. To hear him speak, no one could possibly have guessed that at the time of Diana’s death, she was alienated from her brother. This may or may not have been as a consequence of his inability to relate to her fears and warnings of her impending death. Her psychological disposition definitely seems to have drawn a wedge between them.

Quote: He wrote to her, saying she had mental problems and was manipulative and deceitful, a newspaper reported Wednesday. Spencer, speaking on NBC's "Today" show, said the letter, published in excerpts from a new book, was being seen out of context and that he "adored Diana". Reference:http://www.greatdreams.com/diana4.htm

He also said that he did not believe her fatal 1997 car crash was planned, despite a letter, published in the same book, in which Diana said she feared someone would tamper with her car's brakes. Both letters are from "A Royal

Duty," a forthcoming book by Diana's former butler Paul Burrell, which is being excerpted in the Daily Mirror newspaper. End of Quote.

Section 21. The letter known to the court as "the Mischon note" and the disbelief of Diana's family.

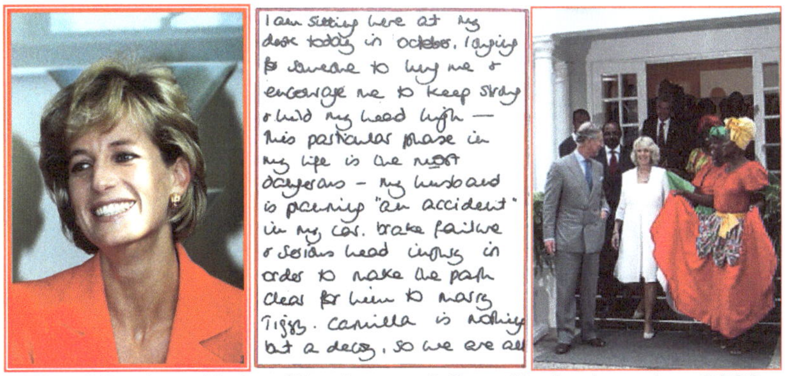
I am sitting here at my desk today in October, longing for someone to hug me & encourage me to keep strong & hold my head high — this particular phase in my life is the most dangerous — my husband is planning 'an accident' in my car, brake failure & serious head injury in order to make the path clear for him to marry Tiggy. Camilla is nothing but a decoy, so we are all

The Mischon Note. Charles and Camilla in Jamaica 2008

"My family and I are absolutely certain that we've never seen any evidence of that whatsoever," Spencer said, speaking to NBC from Toronto. As for Diana's fears, he said, "I do think it's just a horrible coincidence rather than actually tied in with reality."

Asked if Diana's fears were justified, Spencer said she had spoken to him about eavesdroppers and having her private quarters bugged. "I think paranoid is a very strong word, but I think using it in the common way - meaning very, very concerned about yourself - yes, she was at time," Spencer said. The letter attributed to Spencer and published in the Daily Mirror said, "I know how manipulation and deceit are parts of the illness. I hope you are getting treatment for your mental problem." End of Quote.

In the letter she allegedly wrote: Quote: "I am sitting here at my desk today in October, longing for someone to hug me and encourage me to keep strong and hold my head high. This particular phase in my life is the most dangerous - my husband is planning 'an accident' in my car, brake failure and serious head injury in order to make the path clear for him to marry Tiggy. Camilla is nothing but a decoy, so we are all being used by the man in every sense of the word." The letter has been shown at the inquest at the Royal Courts of Justice in London to witnesses who have been challenged over their assertions that the Princess did not fear for her safety. End of Quote.

Badge of the House of Windsor

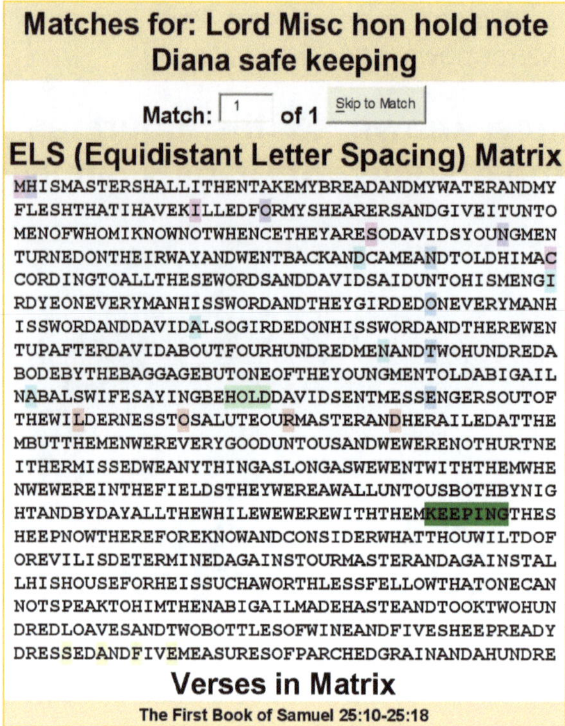

Matches for: Lord Misc hon hold note Diana safe keeping

Match: 1 of 1 Skip to Match

ELS (Equidistant Letter Spacing) Matrix

```
MHISMASTERSHALLITHENTAKEMYBREADANDMYWATERANDMY
FLESHTHATIHAVEKILLEDFORMYSHEARERSANDGIVEITUNTO
MENOFWHOMIKNOWNOTWHENCETHEYARESODAVIDSYOUNGMEN
TURNEDONTHEIRWAYANDWENTBACKANDCAMEANDTOLDHIMAC
CORDINGTOALLTHESEWORDSANDDAVIDSAIDUNTOHISMENGI
RDYEONEVERYMANHISSWORDANDTHEYGIRDEDONEVERYMANH
ISSWORDANDDAVIDALSOGIRDEDONHISSWORDANDTHEREWEN
TUPAFTERDAVIDABOUTFOURHUNDREDMENANDTWOHUNDREDA
BODEBYTHEBAGGAGEBUTONEOFTHEYOUNGMENTOLDABIGAIL
NABALSWIFESAYINGBEHOLDDAVIDSENTMESSENGERSOUTOF
THEWILDERNESSTOSALUTEOURMASTERANDHERAILEDATTHE
MBUTTHEMENWEREVERYGOODUNTOUSANDWEWERENOTHURTNE
ITHERMISSEDWEANYTHINGASLONGASWEWENTWITHTHEMWHE
NWEWEREINTHEFIELDSTHEYWEREAWALLUNTOUSBOTHBYNIG
HTANDBYDAYALLTHEWHILEWEWEREWITHTHEMKEEPINGTHES
HEEPNOWTHEREFOREKNOWANDCONSIDERWHATTHOUWILTDOF
OREVILISDETERMINEDAGAINSTOURMASTERANDAGAINSTAL
LHISHOUSEFORHEISSUCHAWORTHLESSFELLOWTHATONECAN
NOTSPEAKTOHIMTHENABIGAILMADEHASTEANDTOOKTWOHUN
DREDLOAVESANDTWOBOTTLESOFWINEANDFIVESHEEPREADY
DRESSEDANDFIVEMEASURESOFPARCHEDGRAINANDAHUNDRE
```

Verses in Matrix

The First Book of Samuel 25:10-25:18

Section 22. A French court is said to have reopened Princess Diana's death probe.

Quote. The Times of India Fri, 22 Jul 2011 09:55 CDT London. A French court is seeking to question two former top British police officials over allegations that they withheld "crucial evidence" about the Paris car crash which killed her in 1997, a media report said. French Judge Gerard Caddeo wants to interrogate former Scotland Yard chief Lord Condon and ex-Assistant Commissioner of Metropolitan Police Sir David Veness why they had failed to disclose the existence of a note in which the late Princess of Wales predicted her assassination, the Daily Express said.

Under French law, "removing or concealing" evidence, which could "facilitate the discovery of a crime", is punishable by three to five years in jail, it said. In fact, the note, taken by Diana's lawyer Lord Mishcon, was handed to the officers a few months after the 1997 Paris tunnel crash which also claimed the lives of her boyfriend Dodi Al-Fayed.

The lawyer's document records the line: "Efforts would be made if not to get rid of her (be it by some accident in her car, such as a pre-prepared brake failure or whatever)... at least to see that she was so injured or damaged as to be declared unbalanced." But it was more than three years later before it emerged that UK officers had locked the note in Lord Condon's safe at Scotland Yard. When Lord Condon stood down as the Met Police chief, his successor too kept its existence a secret.

A source close to the Paris probe said: "Everyone assumed that with the end of the British inquest, the investigation into Diana and Dodi's deaths was closed. But this has blown it wide open again. Judge Caddeo is determined to get to the bottom of what happened with the Mishcon note." End of quote.

In doing a search to see if I could establish veracity of the letter, in order to get a definitive "Diana", I had to give detail about her that would set her apart from others with the same name, otherwise "Diana" could be anybody, not necessarily the Princess. The word "letter" did not come up, but when I tried the word "note" it did.

As I was writing this down, I noticed, and was reminded, from the paragraph above, that Diana's letter to Paul has been referred to in court as "the Mishcon note". This in my view adds strength to the findings, and if The Bible Code is true, then the note is also true.

Section 23. Interview with Richard Tomlinson ex MI6 agent.

Quote: (New Idea). An explosive interview with secret agent Richard Tomlinson (New Zealand born) that reveals the inside on how the Princess was set up. After months of cloak-and-dagger spying, arrests and claims of dirty tricks against him, former British spy Richard Tomlinson has lifted the lid on the sinister world of MI6 - the super secret agency he once worked for and which he now accuses of hounding and plotting to kill Princess Diana.

Regarded as Britain's No. 1 enemy by royalists, Richard believes that only by going public with what he knows is he safe from becoming MI6's next victim. In an exclusive interview with New Idea in his Geneva hideaway he details, for the first time, what he secretly told French judge Herve Stephan, the man heading the inquest into Diana's death. The results of Judge Stephan's investigation into the Paris car crash, which killed the Princess, her lover Dodi Fayed and Ritz security boss Henri Paul a year ago, are not expected to be released for some months.

Richard claims that Diana was spied on for 15 years until the night she died - and that one of the Paris paparazzi tailing her was, in reality, an MI6 agent. He tells a shocking tale of a spy organization that hounded the Princess every day of her public life. He says: "We were given a standing instruction to keep her in our sights. Even in the early days when she thought she could drive herself to visit friends, she was being followed everywhere - and not just by one car. As she turned out of her gates, up to three cars would move in behind her. The first would then head off to let the second take up the tail. When it peeled off, the third would take over."

At her favourite London restaurant, San Lorenzo, every word the Princess uttered was recorded. Richard says: "No one, including the owner, knew the place was wired - and that the couple at the next table were listening to her every word. I don't think she had a private phone call in years. We had films and tapes of everything she was doing and everyone she was seeing."

Richard also told the inquiry that MI6 renegades, fanatically loyal to the Crown, viewed Diana as a 'loose cannon' - and would stop at nothing to eliminate her threat to eclipse the House of Windsor.

His knowledge of MI6's Diana operation made him a dangerous, high-risk threat to the organization after Diana's crash. Richard was picked up by police just three weeks after her death. He was handcuffed and interrogated. "I knew that Diana's death had MI6's dirty fingerprints all over it," he says. "I was determined to speak out and MI6 was just as determined to silence me..."

He also told the French inquiry:-

Ritz security boss Henri Paul, who drove the death car, was an MI6 informer paid to spy on Diana and Dodi.

The Diana crash was chillingly similar to a previous MI6 plot. That plot was to assassinate the Serbian leader Slobodan Milosevic in Geneva using a powerful laser strobe light - similar to that described by witnesses to the Paris crash - to blind the driver.

Reference:http://www.dailymail.co.uk/news/article-514147/Assassin-murdered-Diana-plan-mirroring-Milosevic-plot-admits-spy.html#ixzz1ru6sTBKw

Richard claims MI6 dirty tricks experts have also perfected transmitters to knock out a car's electronics while it is on the move - even causing the airbags to inflate, blinding the driver. "I was shown a document proposing an assassination of President Milosevic of Serbia because of his involvement with weapons shipments to Bosnia. The plan was to use a strobe light to blind his driver as he went into a road tunnel in Geneva. When I heard witnesses in Paris talk about a bright flash before Diana's car crash, it made sense.

A tunnel is a perfect place for an assassination, with fewer witnesses. The Paris tunnel is also ideal because there are no crash rails along the central pillars, so it's a death trap. I also know that one of the paparazzi that night was working for MI6, too. He'd have been on a fast motorbike."

"They'd also have at their disposal sophisticated radio transmitters that would knock out the electronics of the limo at the press of a button, causing the airbags to inflate. Imagine - the driver flying into the tunnel at 90mph (145kph), picking up speed as he dips down, inches from those central concrete pillars. Suddenly, the car dies as the electronics fail. A blinding flash of light is shot into his eyes. Then, the airbag blows up in his face. Instant disaster..."

His claims about MI6 involvement in Diana's death will shock the world. "There's an arrogant faction inside MI6, part of the Eton/Oxford/Guards clique, who see themselves literally as defenders of the realm - and for them, that means the royals."

"When Di broke up with Charles, she immediately became the enemy. When she started a romance with Dodi Fayed, that raised an even more terrifying spectre. What if she'd married him and turned Muslim? What if they'd had children? The thought of Prince William, the future King of England, with a brown-skinned Muslim half-brother or sister was the worst possible scenario for them. In their eyes, Diana would single-handedly destroy the fabric of the nation they were pledged to defend." End of Quote.

(References: New Idea Magazine) http://koeln.ccc.de/archiv/drt/MI6dir.htm

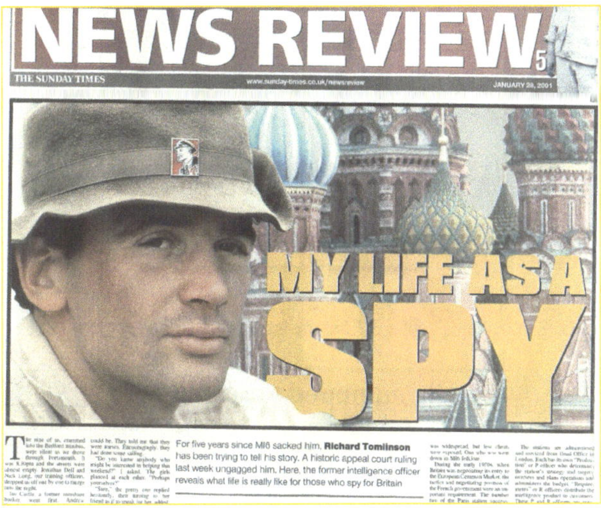

NEWS REVIEW 5

THE SUNDAY TIMES www.sunday-times.co.uk/newsreview JANUARY 28, 2001

MY LIFE AS A SPY

For five years since MI6 sacked him, **Richard Tomlinson** has been trying to tell his story. A historic appeal court ruling last week ungagged him. Here, the former intelligence officer reveals what life is really like for those who spy for Britain

Tomlinson, former MI6 operative, whose explosive revelations about alleged SIS (MI6) involvement in Diana's and Dodi's death sparked controversy and astonished the general public.

Reference:http://www.mi5.com/security/mi6org/tomlinson.htm

Matches for: Alma Pont SIS MOD flash light Henri eyes Diana car

Match: 1 of 1 Skip to Match

ELS (Equidistant Letter Spacing) Matrix

```
COMMANDMENTSWITHTHEESOASTOINCLINETHINEEARUNTOW
ISDOMANDAPPLYTHYHEARTTOUNDERSTANDINGYEAIFTHOUC
RYAFTERDISCERNMENTANDLIFTUPTHYVOICEFORUNDERSTA
NDINGIFTHOUSEEKHERASSILVERANDSEARCHFORHERASFOR
HIDTREASURESTHENSHALTTHOUUNDERSTANDTHEFEAROFJE
HOVAHANDFINDTHEKNOWLEDGEOFGODFORJEHOVAHGIVETHW
ISDOMOUTOFHISMOUTHCOMETHKNOWLEDGEANDUNDERSTAND
INGHELAYETHUPSOUNDWISDOMFORTHEUPRIGHTHEISASHIE
LDTOTHEMTHATWALKININTEGRITYTHATHEMAYGUARDTHEPA
THSOFJUSTICEANDPRESERVETHEWAYOFHISSAINTSTHENSH
ALTTHOUUNDERSTANDRIGHTEOUSNESSANDJUSTICEANDEQU
ITYYEAEVERYGOODPATHFORWISDOMSHALLENTERINTOTHYH
EARTANDKNOWLEDGESHALLBEPLEASANTUNTOTHYSOULDISC
RETIONSHALLWATCHOVERTHEEUNDERSTANDINGSHALLKEEP
THEETODELIVERTHEEFROMTHEWAYOFEVILFROMTHEMENTHA
TSPEAKPERVERSETHINGSWHOFORSAKETHEPATHSOFUPRIGH
TNESSTOWALKINTHEWAYSOFDARKNESSWHOREJOICETODOEV
ILANDDELIGHTINTHEPERVERSENESSOFEVILWHOARECROOK
EDINTHEIRWAYSANDWAYWARDINTHEIRPATHSTODELIVERTH
EEFROMTHESTRANGEWOMANEVENFROMTHEFOREIGNERTHATF
```

Verses in Matrix

The Proverbs 2:1-2:14

Princess Diana was photographed holding a man's hand and it shocked the world. What caused a sensation about this particular display of affection was that he had AIDS. In 1987, when Diana was 26 years old, when public hysteria about the condition was rife, such a public and familiar gesture was almost unheard of from any high-profile person, particularly a member of the royal family. Royals shook hands with the public wearing white cotton gloves. But Diana not only refused to wear gloves, she held dying people in her arms, making unpopular causes front-page news.

Matches for: Tom linson right about Andan son MOD SIS Diana die

Match: 1 of 8 Skip to Match

ELS (Equidistant Letter Spacing) Matrix

```
DASLYDDAWASNIGHUNTOJOPPATHEDISCIPLESHEARINGTHA
TPETERWASTHERESENTTWOMENUNTOHIMENTREATINGHIMDE
LAYNOTTOCOMEONUNTOUSANDPETERAROSEANDWENTWITHTH
EMANDWHENHEWASCOMETHEYBROUGHTHIMINTOTHEUPPERCH
AMBERANDALLTHEWIDOWSSTOODBYHIMWEEPINGANDSHOWIN
GTHECOATSANDGARMENTSWHICHDORCASMADEWHILESHEWAS
WITHTHEMBUTPETERPUTTHEMALLFORTHANDKNEELEDDOWNA
NDPRAYEDANDTURNINGTOTHEBODYHESAIDTABITHAARISEA
NDSHEOPENEDHEREYESANDWHENSHESAWPETERSHESATUPAN
DHEGAVEHERHISHANDANDRAISEDHERUPANDCALLINGTHESA
INTSANDWIDOWSHEPRESENTEDHERALIVEANDITBECAMEKNO
WNTHROUGHOUTALLJOPPAANDMANYBELIEVEDONTHELORDAN
DITCAMETOPASSTHATHEABODEMANYDAYSINJOPPAWITHONE
SIMONATANNERNOWTHEREWASACERTAINMANINCAESAREACO
RNELIUSBYNAMEACENTURIONOFTHEBANDCALLEDTHEITALI
ANBANDADEVOUTMANANDONETHATFEAREDGODWITHALLHISH
OUSEWHOGAVEMUCHALMSTOTHEPEOPLEANDPRAYEDTOGODAL
WAYSHESAWINAVISIONOPENLYASITWEREABOUTTHENINTHH
OUROFTHEDAYANANGELOFGODCOMINGINUNTOHIMANDSAYIN
GTOHIMCORNELIUSANDHEFASTENINGHISEYESUPONHIMAND
BEINGAFFRIGHTEDSAIDWHATISITLORDANDHESAIDUNTOHI
MTHYPRAYERSANDTHINEALMSAREGONEUPFORAMEMORIALBE
FOREGODANDNOWSENDMENTOJOPPAANDFETCHONESIMONWHO
ISSURNAMEDPETERHELODGETHWITHONESIMONATANNERWHO
SEHOUSEISBYTHESEASIDEANDWHENTHEANGELTHATSPAKEU
NTOHIMWASDEPARTEDHECALLEDTWOOFHISHOUSEHOLDSERV
ANTSANDADEVOUTSOLDIEROFTHEMTHATWAITEDONHIMCONT
INUALLYANDHAVINGREHEARSEDALLTHINGSUNTOTHEMHESE
```

Verses in Matrix

The Acts of the Apostles 9:38-10:8

These photographs show Diana as a child, perhaps exhibiting the frustration of a girl separated from her mother, forced to see her only on holidays. The itinerancy of her familial lifestyle seems somehow to be mirrored in these family snapshots.

Diana with a young Aids patient, using her high profile to change the public perception of this cruel disease that is no respecter of persons and does not discriminate between babies and children or adults.

Section 24. What of the claims that Diana's seatbelt was tampered with by intelligence agent/s to ensure she received extensive head injuries, or worse, did not survive the impact?

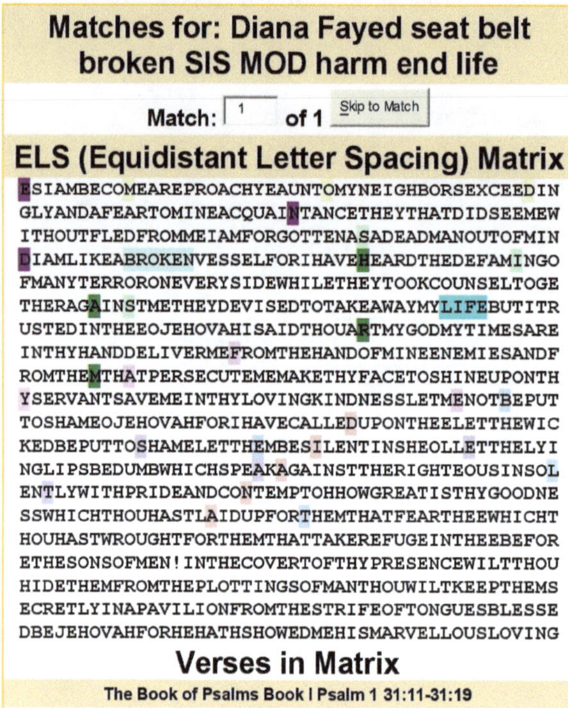

Matches for: Diana Fayed seat belt broken SIS MOD harm end life

Match: 1 of 1 Skip to Match

ELS (Equidistant Letter Spacing) Matrix

ESIAMBECOMEAREPROACHYEAUNTOMYNEIGHBORSEXCEEDIN
GLYANDAFEARTOMINEACQUAINTANCETHEYTHATDIDSEEMEW
ITHOUTFLEDFROMMEIAMFORGOTTENASADEADMANOUTOFMIN
DIAMLIKEABROKENVESSELFORIHAVEHEARDTHEDEFAMINGO
FMANYTERRORONEVERYSIDEWHILETHEYTOOKCOUNSELTOGE
THERAGAINSTMETHEYDEVISEDTOTAKEAWAYMYLIFEBUTITR
USTEDINTHEEOJEHOVAHISAIDTHOUARTMYGODMYTIMESARE
INTHYHANDDELIVERMEFROMTHEHANDOFMINEENEMIESANDF
ROMTHEMTHATPERSECUTEMEMAKETHYFACETOSHINEUPONTH
YSERVANTSAVEMEINTHYLOVINGKINDNESSLETMENOTBEPUT
TOSHAMEOJEHOVAHFORIHAVECALLEDUPONTHEELETTHEWIC
KEDBEPUTTOSHAMELETTHEMBESILENTINSHEOLLETTHELYI
NGLIPSBEDUMBWHICHSPEAKAGAINSTTHERIGHTEOUSINSOL
ENTLYWITHPRIDEANDCONTEMPTOHHOWGREATISTHYGOODNE
SSWHICHTHOUHASTLAIDUPFORTHEMTHATFEARTHEEWHICHT
HOUHASTWROUGHTFORTHEMTHATTAKEREFUGEINTHEEBEFOR
ETHESONSOFMEN!INTHECOVERTOFTHYPRESENCEWILTTHOU
HIDETHEMFROMTHEPLOTTINGSOFMANTHOUWILTKEEPTHEMS
ECRETLYINAPAVILIONFROMTHESTRIFEOFTONGUESBLESSE
DBEJEHOVAHFORHEHATHSHOWEDMEHISMARVELLOUSLOVING

Verses in Matrix

The Book of Psalms Book I Psalm 1 31:11-31:19

In the above Matrix, two of the words found "broken" and "life" are in the ordinary text of The Bible. This must certainly arouse doubts about the strength/ veracity of these findings, which are decidedly weakened. Once again, an exclamation mark, which should not be there, appears in the fourth bottom line of the computer text, but does not impinge on the equidistant letter findings within the verses.

Quote: A study of pictures taken by photographers during the course of the night showed that Trevor Rees Jones did not wear a seat belt when the Mercedes left the Paris Ritz hotel on the night in question. But later pictures showed Rees Jones wearing the belt shortly before the August 31 crash, which also killed Diana's companion Dodi Fayed and the driver of the car. Police believe the last minute action saved his life. End of Quote.

Reference:http://articles.cnn.com/1997-09 21/world/9709_21_diana.investigation_1_dodi-fayed-lalma-seat-belts?_s=PM:WORLD

Matches for: Diana Fayed seat belt broken SIS MOD lies not true

Match: 1 **of 1** Skip to Match

ELS (Equidistant Letter Spacing) Matrix

```
MEAREPROACHYEAUNTOMYNEIGHBORSEXCEEDINGLYANDAFE
ARTOMINEACQUAINTANCETHEYTHATDIDSEEMEWITHOUTFLE
DFROMMEIAMFORGOTTENASADEADMANOUTOFMINDIAMLIKEA
BROKENVESSELFORIHAVEHEARDTHEDEFAMINGOFMANYTERR
ORONEVERYSIDEWHILETHEYTOOKCOUNSELTOGETHERAGAIN
STMETHEYDEVISEDTOTAKEAWAYMYLIFEBUTITRUSTEDINTH
EEOJEHOVAHISAIDTHOUARTMYGODMYTIMESAREINTHYHAND
DELIVERMEFROMTHEHANDOFMINEENEMIESANDFROMTHEMTH
ATPERSECUTEMEMAKETHYFACETOSHINEUPONTHYSERVANTS
AVEMEINTHYLOVINGKINDNESSLETMENOTBEPUTTOSHAMEOJ
EHOVAHFORIHAVECALLEDUPONTHEELETTHEWICKEDBEPUTT
OSHAMELETTHEMBESILENTINSHEOLLETTHELYINGLIPSBED
UMBWHICHSPEAKAGAINSTTHERIGHTEOUSINSOLENTLYWITH
PRIDEANDCONTEMPTOHHOWGREATISTHYGOODNESSWHICHTH
OUHASTLAIDUPFORTHEMTHATFEARTHEEWHICHTHOUHASTWR
OUGHTFORTHEMTHATTAKEREFUGEINTHEEBEFORETHESONSO
FMEN!INTHECOVERTOFTHYPRESENCEWILTTHOUHIDETHEMF
ROMTHEPLOTTINGSOFMANTHOUWILTKEEPTHEMSECRETLYIN
APAVILIONFROMTHESTRIFEOFTONGUESBLESSEDBEJEHOVA
HFORHEHATHSHOWEDMEHISMARVELLOUSLOVINGKINDNESSI
```

Verses in Matrix

The Book of Psalms Book I Psalm 1 31:11-31:19

Once again, the above matrix is ambiguous and therefore open to interpretation. Are SIS and MOD lying when denying their involvement, or are they to be exonerated from involvement? It's up to you to decide, based on an overall assessment of the matrices.

I was born and raised in Australia, a Commonwealth country. I was five years old when Elizabeth II ascended the throne in 1953. My era, until recent times, has been marked by respect for the sovereign. Naturally, I am loath to believe any royal involvement. It goes against my grain.

Furthermore, if I am to believe my late mother, Lillian, her family was related to the Queen Mother through cousinship via paternal heritage. I was told my ancestor was Lady Sophia Jeffreys. All the more reason for me to rail against any negative findings. Australia is now well on the way to becoming a republic. If it should be proven in the courts that there was any kind of royal influence involved in Diana's death, then it is my guess that Britain is sure to follow, with that country going down the same path in a quantum leap, as public opinion radically turns the tide in favour of republicanism.

Once again, the above matrix is subject to ambiguity. Are they lying when they tell us it is not true that an intelligence agent tampered with the seat belt/s of the Fayed vehicle? Or is it a lie that an agent was instructed to tamper with the car's seat belt/s, rendering at least one of them (i.e. Diana's) inoperative?

The Duke of Edinburgh has always seemed to be a thoroughly likeable fellow, despite rumours of his alleged marital indiscretions, a loving and supportive husband to HRH Elizabeth. His sense of humour is legendary around the world. Charles I have always admired, not only as a devoted father to his sons, but also as a man among men in this modern age: keen conservationist, organic gardener, and humanitarian.

Given that the charges are baseless, we can only imagine the suffering the royal family, particularly Philip, who was at the centre of Dodi's father's accusations, have had to endure on top of the tragic death of Diana. However, just because there has been no (immediate familial) royal involvement in Diana's death, that does not mean there has been no conspiracy to end her life, or seriously injure and disfigure her. That may not let MOD, SIS and the French Intel off the hook. It may not let other royals, further from the throne, off the hook either. Diana may have received warnings from "a sympathetic insider," as has been claimed. So, we need to check it out.

You will notice in the matrix above, that there is an exclamation mark in the Biblical verse, fifth bottom line. I think it perhaps should not be there. If researchers are going to come up with findings, computer programmers leave their systems open to criticism and thus our findings, if they leave punctuation marks in the text. Fortunately, its presence there does not impinge on these results at all.

The text was allegedly given to Moses as unbroken letters, without punctuation. For the first time in history, if it was indeed given this way, the manner in which it was transmitted makes sense, now that the computer has been invented, and we are able for the first time ever, to search for matrices within the Biblical text.

It is been suggested that had Diana and Dodi been wearing seatbelts, one or both of them might have survived the impact. The reasoning behind this is that the only person wearing a seatbelt at the time of the impact, survived. Although the front of the car is extensively damaged, the rear seat section looks intact. However, force of the impact is what counts in an accident, and injuries are likely to be fatal in a high impact collision, (as this one was) especially if the occupants are not wearing seat belts.

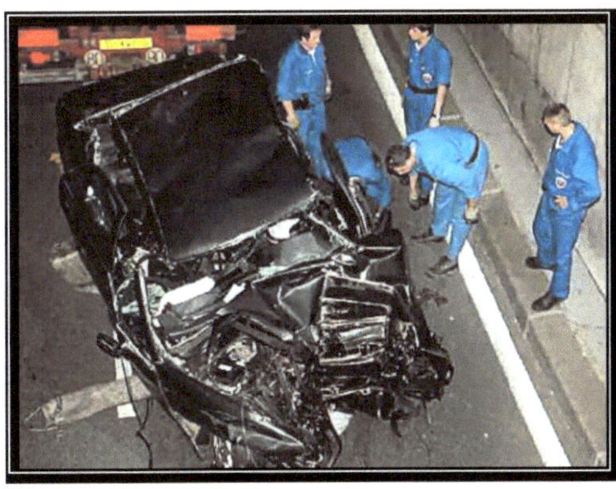

Matches for: MOD SIS not solely plot Fiat motor car tip

Match: 1 of 4 Skip to Match

ELS (Equidistant Letter Spacing) Matrix

```
NHEWHOISINHEAVENNEITHERBEYECALLEDMASTERSFORONE
ISYOURMASTEREVENTHECHRISTBUTHETHATISGREATESTAM
ONGYOUSHALLBEYOURSERVANTANDWHOSOEVERSHALLEXALT
HIMSELFSHALLBEHUMBLEDANDWHOSOEVERSHALLHUMBLEHI
MSELFSHALLBEEXALTEDBUTWOEUNTOYOUSCRIBESANDPHAR
ISEESHYPOCRITES!BECAUSEYESHUTTHEKINGDOMOFHEAVE
NAGAINSTMENFORYEENTERNOTINYOURSELVESNEITHERSUF
FERYETHEMTHATAREENTERINGINTOENTERWOEUNTOYOUSCR
IBESANDPHARISEESHYPOCRITES!FORYEDEVOURWIDOWSHO
USESEVENWHILEFORAPRETENCEYEMAKELONGPRAYERSTHER
EFOREYESHALLRECEIVEGREATERCONDEMNATIONWOEUNTOY
OUSCRIBESANDPHARISEESHYPOCRITES!FORYECOMPASSSE
AANDLANDTOMAKEONEPROSELYTEANDWHENHEISBECOMESOY
EMAKEHIMTWOFOLDMOREASONOFHELLTHANYOURSELVESWOE
UNTOYOUYEBLINDGUIDESTHATSAYWHOSOEVERSHALLSWEAR
BYTHETEMPLEITISNOTHINGBUTWHOSOEVERSHALLSWEARBY
THEGOLDOFTHETEMPLEHEISADEBTORYEFOOLSANDBLINDFO
RWHICHISGREATERTHEGOLDORTHETEMPLETHATHATHSANCT
IFIEDTHEGOLDANDWHOSOEVERSHALLSWEARBYTHEALTARIT
ISNOTHINGBUTWHOSOEVERSHALLSWEARBYTHEGIFTTHATIS
UPONITHEISADEBTORYEBLINDFORWHICHISGREATERTHEGI
FTORTHEALTARTHATSANCTIFIETHTHEGIFTTHETHEREFORET
HATSWEARETHBYTHEALTARSWEARETHBYITANDBYALLTHING
STHEREONANDHETHATSWEARETHBYTHETEMPLESWEARETHBY
ITANDBYHIMTHATDWELLETHTHEREINANDHETHATSWEARETH
BYTHEHEAVENSWEARETHBYTHETHRONEOFGODANDBYHIMTHA
```

Verses in Matrix

The Gospel According to St. Matthew 23:9-23:22

Matches for: MOD SIS not solely plan Diana die road Fiat tip

Match: 1 of 1 Skip to Match

ELS (Equidistant Letter Spacing) Matrix

```
READSHALLBEEATENANDUPONTHATDAYSHALLTHEPRINCEPR
EPAREFORHIMSELFANDFORALLTHEPEOPLEOFTHELANDABUL
LOCKFORASINOFFERINGANDTHESEVENDAYSOFTHEFEASTHE
SHALLPREPAREABURNTOFFERINGTOJEHOVAHSEVENBULLOC
KSANDSEVENRAMSWITHOUTBLEMISHDAILYTHESEVENDAYSA
NDAHEGOATDAILYFORASINOFFERINGANDHESHALLPREPARE
AMEALOFFERINGANEPHAHFORABULLOCKANDANEPHAHFORAR
AMANDAHINOFOILTOANEPHAHINTHESEVENTHMONTHINTHEF
IFTEENTHDAYOFTHEMONTHINTHEFEASTSHALLHEDOTHELIK
ETHESEVENDAYSACCORDINGTOTHESINOFFERINGACCORDIN
GTOTHEBURNTOFFERINGANDACCORDINGTOTHEMEALOFFERI
NGANDACCORDINGTOTHEOILTHUSSAITHTHELORDJEHOVAHT
HEGATEOFTHEINNERCOURTTHATLOOKETHTOWARDTHEEASTS
HALLBESHUTTHESIXWORKINGDAYSBUTONTHESABBATHDAYI
TSHALLBEOPENEDANDONTHEDAYOFTHENEWMOONITSHALLBE
OPENEDANDTHEPRINCESHALLENTERBYTHEWAYOFTHEPORCH
OFTHEGATEWITHOUTANDSHALLSTANDBYTHEPOSTOFTHEGAT
EANDTHEPRIESTSSHALLPREPAREHISBURNTOFFERINGANDH
ISPEACEOFFERINGSANDHESHALLWORSHIPATTHETHRESHOL
DOFTHEGATETHENHESHALLGOFORTHBUTTHEGATESHALLNOT
BESHUTUNTILTHEEVENINGANDTHEPEOPLEOFTHELANDSHAL
LWORSHIPATTHEDOOROFTHATGATEBEFOREJEHOVAHONTHES
ABBATHSANDONTHENEWMOONSANDTHEBURNTOFFERINGTHAT
THEPRINCESHALLOFFERUNTOJEHOVAHSHALLBEONTHESABB
ATHDAYSIXLAMBSWITHOUTBLEMISHANDARAMWITHOUTBLEM
ISHANDTHEMEALOFFERINGSHALLBEANEPHAHFORTHERAMAN
DTHEMEALOFFERINGFORTHELAMBSASHEISABLETOGIVEAND
AHINOFOILTOANEPHAHANDONTHEDAYOFTHENEWMOONITSHA
LLBEAYOUNGBULLOCKWITHOUTBLEMISHANDSIXLAMBSANDA
RAMTHEYSHALLBEWITHOUTBLEMISHANDHESHALLPREPAREA
```

Verses in Matrix

The Book of the Prophet Ezekiel 45:21-46:6

MOD = Ministry of Defence London.

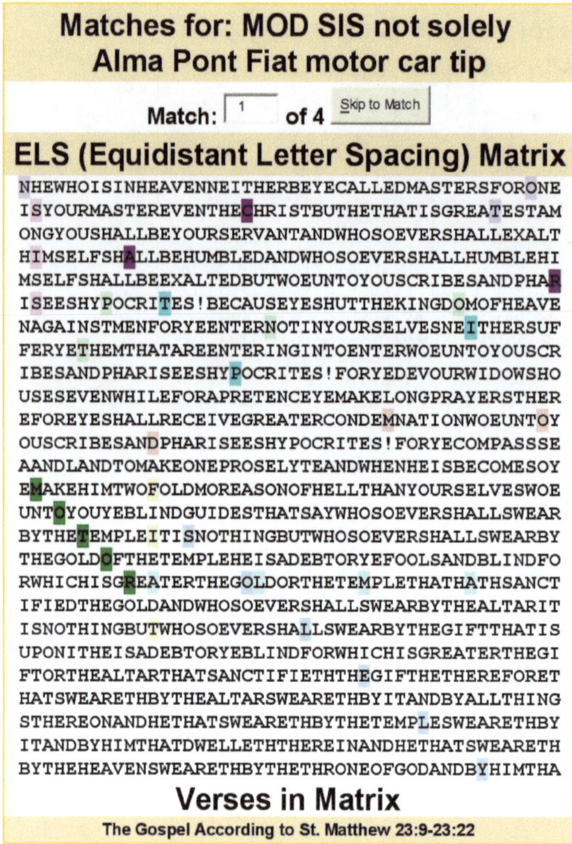

Although the word "not" appears in the last three matrices above, the words "not true" and "false" cannot be added. If The Bible Code is true, we are left with the conclusion that the MOD and SIS were not solely responsible: that is, someone else was involved.

Section 25. French intel, FBI, CIA – their alleged involvement. Quote: French intel chief: detained Islamists a threat.

AP Associated Press By ELAINE GANLEY | Associated Press – Sat, Mar 31, 2012 PARIS (AP) The head of France's counterespionage agency says that Islamist radicals detained in a sweep around the country had a stash of arms, did physical training and may have been preparing a kidnapping. End of Quote.

As you can see from the above newspaper headline, the word "intel" is commonly used by the English media in conjunction with the French Intelligence. Note the newspaper heading does not say "Detained Islamist extremists a threat." I think that those of the Islamic faith might have been entitled to take offence from that newspaper heading, and it may reveal something about the way these incidents are sometimes worded in reportage.

After all, Christians would take offence if a newspaper heading was worded "Detained Christians a threat", and there are plenty of dangerous extremists among them. Take the Ku Klux Klan for instance.

Those who profess that the Muslim faith is not discriminated against in predominantly Christian society had better take a closer look. This kind of thing gives weight to Tomlinson's claim that the British establishment was very uneasy indeed about Princess Diana's alleged forthcoming union with Dodi Fayed and hence the Muslim faith.

From the matrix below, the words "not true" could be said to apply only to "MOD" and "SIS", implicating the French or other intelligence agencies around the world in her death.

Matches for: MOD SIS not solely cause Diana die other did also

Match: 1 of 1 Skip to Match

ELS (Equidistant Letter Spacing) Matrix

```
SEYEKNOWHIMWHOISFROMTHEBEGINNINGIHAVEWRITTENUN
TOYOUYOUNGMENBECAUSEYEARESTRONGANDTHEWORDOFGOD
ABIDETHINYOUANDYEHAVEOVERCOMETHEEVILONELOVENOT
THEWORLDNEITHERTHETHINGSTHATAREINTHEWORLDIFANY
MANLOVETHEWORLDTHELOVEOFTHEFATHERISNOTINHIMFOR
ALLTHATISINTHEWORLDTHELUSTOFTHEFLESHANDTHELUST
OFTHEEYESANDTHEVAINGLORYOFLIFEISNOTOFTHEFATHER
BUTISOFTHEWORLDANDTHEWORLDPASSETHAWAYANDTHELUS
TTHEREOFBUTHETHATDOETHTHEWILLOFGODABIDETHFOREV
ERLITTLECHILDRENITISTHELASTHOURANDASYEHEARDTHA
TANTICHRISTCOMETHEVENNOWHAVETHEREARISENMANYANT
ICHRISTSWHEREBYWEKNOWTHATITISTHELASTHOURTHEYWE
NTOUTFROMUSBUTTHEYWERENOTOFUSFORIFTHEYHADBEENO
FUSTHEYWOULDHAVECONTINUEDWITHUSBUTTHEYWENTOUTT
HATTHEYMIGHTBEMADEMANIFESTTHATTHEYALLARENOTOFU
SANDYEHAVEANANOINTINGFROMTHEHOLYONEANDYEKNOWAL
LTHETHINGSIHAVENOTWRITTENUNTOYOUBECAUSEYEKNOWN
OTTHETRUTHBUTBECAUSEYEKNOWITANDBECAUSENOLIEISO
FTHETRUTHWHOISTHELIARBUTHETHATDENIETHTHATJESUS
ISTHECHRISTTHISISTHEANTICHRISTEVENHETHATDENIET
HTHEFATHERANDTHESONWHOSOEVERDENIETHTHESONTHESA
MEHATHNOTTHEFATHERHETHATCONFESSETHTHESONHATHTH
EFATHERALSOASFORYOULETTHATABIDEINYOUWHICHYEHEA
RDFROMTHEBEGINNINGIFTHATWHICHYEHEARDFROMTHEBEG
INNINGABIDEINYOUYEALSOSHALLABIDEINTHESONANDINT
HEFATHERANDTHISISTHEPROMISEWHICHHEPROMISEDUSEV
ENTHELIFEETERNALTHESETHINGSHAVEIWRITTENUNTOYOU
CONCERNINGTHEMTHATWOULDLEADYOUASTRAYANDASFORYO
UTHEANOINTINGWHICHYERECEIVEDOFHIMABIDETHINYOUA
```

Verses in Matrix

The First Epistle General of John 2:14-2:27

So, I decided to do a search for "French Intel" alone. However, I could not get a match for "Alma Tunnel" or "Alma Pont" as I have been able to in other searches. This renders the matrix below suspect, in that it does not relate specifically enough to Princess Diana, as an individual.

Diana with a shy and possibly a bit tired, Harry

Matches for: MOD SIS Diana die not true France Intel Fiat tip

Match: 1 of 6 Skip to Match

ELS (Equidistant Letter Spacing) Matrix

```
NPLACEANDBEMOVEDNOMORENEITHERSHALLTHECHILDRENO
FWICKEDNESSWASTETHEMANYMOREASATTHEFIRSTANDASFR
OMTHEDAYTHATICOMMANDEDJUDGESTOBEOVERMYPEOPLEIS
RAELANDIWILLSUBDUEALLTHINEENEMIESMOREOVERITELL
THEETHATJEHOVAHWILLBUILDTHEEAHOUSEANDITSHALLCO
METOPASSWHENTHYDAYSAREFULFILLEDTHATTHOUMUSTGOT
OBEWITHTHYFATHERSTHATIWILLSETUPTHYSEEDAFTERTHE
EWHOSHALLBEOFTHYSONSANDIWILLESTABLISHHISKINGDO
MHESHALLBUILDMEAHOUSEANDIWILLESTABLISHHISTHRON
EFOREVERIWILLBEHISFATHERANDHESHALLBEMYSONANDIW
ILLNOTTAKEMYLOVINGKINDNESSAWAYFROMHIMASITOOKIT
FROMHIMTHATWASBEFORETHEEBUTIWILLSETTLEHIMINMYH
OUSEANDINMYKINGDOMFOREVERANDHISTHRONESHALLBEES
TABLISHEDFOREVERACCORDINGTOALLTHESEWORDSANDACC
ORDINGTOALLTHISVISIONSODIDNATHANSPEAKUNTODAVID
THENDAVIDTHEKINGWENTINANDSATBEFOREJEHOVAHANDHE
SAIDWHOAMIOJEHOVAHGODANDWHATISMYHOUSETHATTHOUH
ASTBROUGHTMETHUSFARANDTHISWASASMALLTHINGINTHIN
EEYESOGODBUTTHOUHASTSPOKENOFTHYSERVANTSHOUSEFO
RAGREATWHILETOCOMEANDHASTREGARDEDMEACCORDINGTO
```

Verses in Matrix

The First Book of the Chronicles 17:9-17:16

Ambiguously, the matrix above could be interpreted to mean one of three things: (a) the French Intel" did not arrange the accident, thereby eliminating them from responsibility for it; that MOD and SIS did it. Or, (b) it could be thought to mean that MOD and SIS are eliminated from responsibility, as the French Intel arranged it. Or, (c) that none of MOD, SIS or the French Intel were responsible.

Diana as a school girl and as a pre-school teacher, wearing the see-through flower-patterned dress that was the order of the day for young women around the Western world, yet aroused a storm with the press because of her association with Prince Charles as his intended bride .

The results of the matrices below, if the Bible Code is true, would seem to indicate the French intelligence were not involved in any conspiracy to kill Diana or injure her in the car crash in the Alma Tunnel. However, it is worth noting that I was unable to get either the words "Alma" or "Tunnel" included in my results, making them suspicious. The "Diana" in the matrix below, it might be argued, could be any Diana around the world.

Engagement photo. Kate, William's wife, now wears Diana's engagement ring.

Penny Thornton, Diana's astrologer.

However, the match in the matrix below is more definitive, and if The Bible Code is true, it seems to let the French Intel off the hook. Unless, of course, it was not the French intelligence who did the actual tipping of the Mercedes using the white Fiat Uno. They may have been involved in other ways. We have here been looking only at the possibility of their possible involvement in the sabotage of the Mercedes vehicle. The French Intelligence Building is, according to Google directions, eighteen minutes drive from the Pont de l'Alma tunnel where Diana and Dodi died.

Well, there is one other possibility of intelligence involvement in Diana's and Dodi's deaths, and that is the alleged involvement of USA intelligence operatives. So, let's take a look at what The Bible Code has to say about that possibility. .

Insignia of the French Intelligence: "In every place where necessity makes the law."

Section 26. Was the United States FBI and/or CIA involved in Diana's death? US bugged Diana's phone on night of death crash.

If The Bible Code is true - N.B. - the double negation in the matrix that follows this discussion renders the assertion "It's not true that the USA CIA FBI killed Diana" a necessarily false one.

Quote: Mark Townsend and Peter Allen in Paris The Observer, Sunday 10 December 2006. The American secret service was bugging Princess Diana's telephone conversations without the approval of the British security services on the night she died, according to the most comprehensive report on her death, to be published this week.

Among extraordinary details due to emerge in the report by former Metropolitan police commissioner Lord Stevens is the revelation that the US security service was bugging her calls in the hours before she was killed in a car crash in Paris.

In a move that raises fresh questions over transatlantic agreements on intelligence-sharing, the surveillance arm of the US has admitted listening to her conversations as she stayed at the Ritz hotel, but failed to notify MI6.

Stevens is understood to have been assured that the 39 classified documents detailing Diana's final conversations did not reveal anything sinister or contain material that might help explain her death. Scotland Yard's inquiry, published this Thursday, also throws up further intelligence links with the Princess of Wales on the night she died. The driver of the Mercedes, Henri Paul, was in the pay of the French equivalent of M15. Stevens traced £100,000 he had amassed in 14 French bank accounts though no payments have been linked to Diana's death.

Stevens's conclusion is that Diana, her companion Dodi Fayed, and Paul himself died in an accident caused by Paul driving too fast through the Pont de l'Alma underpass in Paris while under the influence of drink. The car was being pursued by photographers at the time.

Tests have confirmed that Paul was more than three times over the French drink-drive limit and was travelling at 'excessive' speed. The inquiry will quash a number of conspiracy theories that have circulated since 31 August 1997, among them that Diana was pregnant. It also found no evidence that the princess was planning to get engaged to Dodi, son of Mohamed Fayed.

The Harrods tycoon believes that Paul's blood samples were swapped to portray him as a drunk in an elaborate cover-up by the establishment to stop Diana marrying Dodi, a Muslim. Stevens is expected to concede that while there was a mix-up it was an accident and that the original French post-mortem which found that Paul was three-times over the French drink-drive limit was correct.

He is also expected to discount the role of the white Fiat Uno which struck Diana's car shortly before the crash, even though British police officers have failed to track down the vehicle which left white paintwork on the black Mercedes. The inquiry will support the findings of the original French accident inquiry in criticising the paparazzi as a possible reason for encouraging Paul to speed. The 'bright light' theory - the claim that the driver was deliberately blinded by a beam immediately before the crash - is also dismissed by Stevens. End of Quote.

Reference: http://www.guardian.co.uk/uk/2006/dec/10/monarchy.peterallen

Under English law, an inquest is required in cases of sudden or unexplained death. The inquests into the deaths of Diana, Princess of Wales, and Dodi Fayed opened on 8th January 2007, with Dame Elizabeth Butler-Sloss acting as Deputy Coroner of the Queen's Household for the Diana inquest and Assistant Deputy Coroner for Surrey in relation to the Fayed inquest.

Reference:http://en.wikipedia.org/wiki/Death_of_Diana,_Princess of_Wales#2007_inquest

Butler-Sloss originally intended to sit without a jury; this decision was later overturned by the High Court, as well as the jurisdiction of the Coroner of the Queen's Household. On 24th April 2007, Butler-Sloss stepped down, saying she lacked the experience required to deal with an inquest with a jury. The role of Coroner for the inquests was transferred to Lord Justice Scott Baker, who formally took up the role on 11th June as Coroner for Inner West London. On 27th July 2007, Baker, following representations for the lawyers of the interested parties, issued a list of issues likely to be raised at the inquest, many of which had been dealt with in great detail by Operation Paget.

The issues identified were:-

- Whether driver error on the part of Henri Paul caused or contributed to the cause of the collision
- Whether Henri Paul's ability to drive was impaired through drink or drugs
- Whether a Fiat Uno or any other vehicle caused or contributed to the collision
- Whether the actions of the Paparazzi caused or contributed to the cause of the collision
- Whether the road/tunnel layout and construction were inherently dangerous and, if so, whether this contributed to the collision
- Whether any bright/flashing lights contributed to or caused the collision and, if so, their source
- Whose decision it was that the Princess of Wales and Dodi Al Fayed should leave from the rear entrance to the Ritz and that Henri Paul should drive the vehicle
- Henri Paul's movements between 7 and 10 pm on 30th August 1997
- The explanation for the money in Henri Paul's possession on 30th August 1997 and in his bank account
- Whether Andanson was in Paris on the night of the collision
- Whether Diana's life would have been saved if she had reached hospital sooner or if her medical treatment had been different
- Whether Diana was pregnant
- Whether Diana and Dodi Al Fayed were about to announce their engagement
- Whether and, if so in what circumstances, the Princess of Wales feared for her life
- The circumstances relating to the purchase of the ring
- The circumstances in which Diana's body was embalmed
- Whether the evidence of Tomlinson throws any light on the collision
- Whether the British or any other security services had any involvement in the collision
- Whether there was anything sinister about (i) the Cherruault burglary or (ii) the disturbance at the Big Pictures agency

- Whether correspondence belonging to the Princess of Wales (including some from Prince Philip) has disappeared, and if so, the circumstances.

The inquests officially began on 2nd October 2007 with the swearing of a jury of six women and five men. Lord Justice Scott Baker delivered a lengthy opening statement giving general instructions to the jury and introducing the evidence. Mohamed Al-Fayed had claimed that his son and Diana were murdered by the Royal Family.

The inquest heard evidence from people connected with Diana and the events leading to her death, including Paul Burrell, Mohamed Al-Fayed, her stepmother, the survivor of the crash, and the former head of MI5. Lord Justice Scott Baker began his summing up to the jury on 31st March 2008. He stated there was "not a shred of evidence" that the Duke of Edinburgh ordered the death of Diana, Princess of Wales, or that the security services organized it. Lord Justice Scott Baker concluded his summing up on Wednesday, 2nd April 2008.

After summing up, the jury retired to consider five verdicts, namely unlawful killing by the negligence of either or both the following vehicles or Henri Paul; accidental death or an open verdict. The jury decided on 7th April 2008 that Diana had been unlawfully killed by the grossly negligent driving of chauffeur Henri Paul and following vehicles.

White Fiat Uno of the type that allegedly hit Diana's and Dodi's black Mercedes.

The painting above left: Alma Parens, by William Adolphe Bouguereau. Alma variously means kind, nourishing, motherly, fostering, the child feeds one's soul, lifts the spirits. The picture on the right shows Diana doing healing meditation and prayer on a child with Aids. That Diana, Queen of Hearts, died in the Alma tunnel, seems a fitting, if tragic, tribute.

Section 27. Was Diana six weeks pregnant?

Less than three weeks after the deaths of Diana and Dodi, rumours began to run rife that Diana was conscious at the scene of the accident and had confided to medic/s around her, that she was six weeks pregnant. The tragic story continued that Diana had watched Dodi die beside her, and realizing both the extent of her injuries and she and her unborn child were soon to follow him to the grave, she had requested that she be buried with her baby's father and intended husband.

Quote: On September 19, 1997, Time quoted Agence France-Presse as saying that Diana may have been six-weeks pregnant at the time of the crash.

On July 15, 1997, The Chicago Tribune reported that Diana was pictured leaning over a yacht and reportedly telling the reporters: "You are going to get a big surprise, you'll see. You are going to get a big surprise with the next thing I do." Was she hinting at becoming pregnant, or getting engaged to Dodi? On September 8th 1997, after the Princess' death, Time magazine said Diana had later denied making the statement.

Frederic Maillez, an emergency worker with SOS Medecins, who was the first medic on the scene of the crash, said that an associate at the crash site told him that Diana was drifting in and out of consciousness, and at one point saying she was six-weeks pregnant while making a rubbing gesture on her belly. Apparently, the doctor now refuses to say how she looked or what she said. Furthermore, Mailliez was quoted as saying that if she told him anything: "I would say it only to her family, her sons, her husband."

At some point in the morning, according to the News of the World (23rd November 1997), Diana had requested that she be buried with Dodi. If this is true, who did Diana tell this to, and why did the Spencer family not honour this final wish of hers? If they were not told about it, then why not? Is this because the royal family would never have allowed it? Remember, it was diplomats from the British embassy in Paris who were the first official representatives to reach the hospital, and they would have reported any details straight back to Buckingham Palace first.

Finally, why was Diana's body left alone in the chapel for almost a week during which no member of the royal family visited her? Is it because, as the Independent on Sunday suggested in an article "Was Diana Murdered?" in 1997: "Her refusal to be cowed, to go quietly, had taken her into new avenues, to campaigning for the abolition of land mines to the fury of the United States government and the shadowy arms community.

She was becoming ever more closely involved with a Muslim, and not just any old Muslim but the son of a man who was, and is, a thorn in the side of the Conservative British establishment. So, she was cut down." (Independent on Sunday, 19th October 1997). End of Quote.

Reference: http://www.fantompowa.net/Flame/diana.htm#Was Diana pregnant six weeks pr

Prince William as a baby, with his Mummy Diana and Daddy Charles. Diana's influence over the children was enormous. They loved her then with a passion, as they still do today. Had she become a Muslim when she married Dodi, the establishment may have feared her sons would convert too. A Muslim King of England? Would that thought have spooked them?

Section 28. Fear of the influence of Jemima Khan on Diana? A (dangerous) precedent set? A request for a Muslim/Christian burial?

Jemima Khan, Diana's close friend and daughter of the English establishment, sets a precedent by converting to Islam in 1995, two years prior to Diana's death in 1997. Diana's close friend Jemima Khan, (nee Goldsmith and daughter of Lady Annabel and Sir James), had become a Muslim when she married Imran Khan, the dashing (Islamic) test cricketer from Pakistan.

Did the British establishment fear Diana would soon follow in her friend Jemima's freshly trodden footsteps? In light of this historic timeline, Tomlinson's claims (that the British establishment feared Diana's conversion to the Muslim faith), don't seem quite so outlandish. If they were to get wind of the fact that Diana was pregnant and about to marry Dodi, how might they have reacted?

The Number 21, ruling Diana's life from birth to age 27, equates with a figure dancing on top of the world: "the hand that rocks the cradle rules the world."

The number 21 is a dancer's number, the number 3, a stage performer. To my knowledge, this was the only time the public ever saw Princess Diana dance on stage. The routine was evidently rehearsed in secret, then sprang as a birthday surprise for Charles. He is reported to have said, "I can't believe how good she was!" I have been unable to find any video of this live-on-stage performance on the Internet, although there are still photographs like those above inserted into Billy Joel's *Uptown Girl* on YouTube.

Even so, from just the black and white stills alone, we can perceive it to be a beautiful performance. It should be replayed for the public to view on a huge screen at any of the celebrations of Diana's life, and there have been many since her death. It is testament to both the repression of Diana as a woman and suppression of her talents as an artist, that this vital component of her life is almost entirely missing, concealed from public view, yet literally thousands of photographs of Diana are readily obtainable anywhere on the Internet.

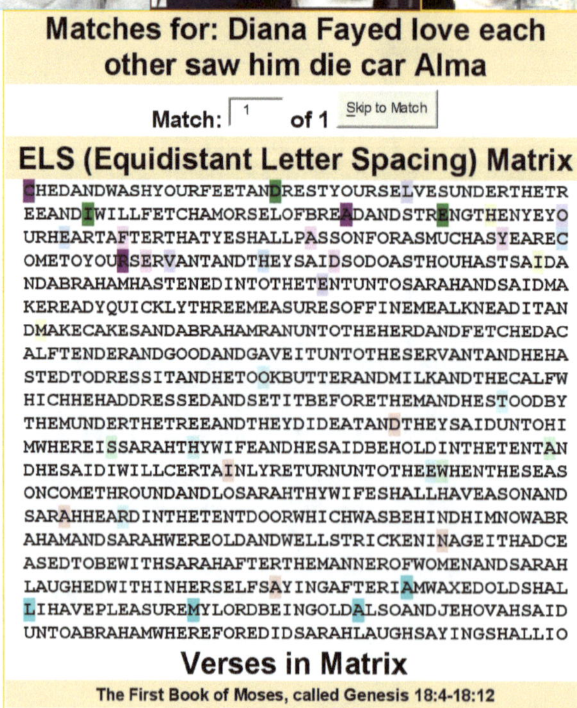

Matches for: Diana asked bury with child father she saw die

Match: 1 of 1 Skip to Match

ELS (Equidistant Letter Spacing) Matrix

```
SENTEMPTYAWAYHEHATHGIVENHELPTOISRAELHISSERVANT
THATHEMIGHTREMEMBERMERCYASHESPAKEUNTOOURFATHER
STOWARDABRAHAMANDHISSEEDFOREVERANDMARYABODEWIT
HHERABOUTTHREEMONTHSANDRETURNEDUNTOHERHOUSENOW
ELISABETHSTIMEWASFULFILLEDTHATSHESHOULDBEDELIV
EREDANDSHEBROUGHTFORTHASONANDHERNEIGHBORSANDHE
RKINSFOLKHEARDTHATTHELORDHADMAGNIFIEDHISMERCYT
OWARDSHERANDTHEYREJOICEDWITHHERANDITCAMETOPASS
ONTHEEIGHTHDAYTHATTHEYCAMETOCIRCUMCISETHECHILD
ANDTHEYWOULDHAVECALLEDHIMZACHARIASAFTERTHENAME
OFTHEFATHERANDHISMOTHERANSWEREDANDSAIDNOTSOBUT
HESHALLBECALLEDJOHNANDTHEYSAIDUNTOHERTHEREISNO
NEOFTHYKINDREDTHATISCALLEDBYTHISNAMEANDTHEYMAD
ESIGNSTOHISFATHERWHATHEWOULDHAVEHIMCALLEDANDHE
ASKEDFORAWRITINGTABLETANDWROTESAYINGHISNAMEISJ
OHNANDTHEYMARVELLEDALLANDHISMOUTHWASOPENEDIMME
DIATELYANDHISTONGUELOOSEDANDHESPAKEBLESSINGGOD
ANDFEARCAMEONALLTHATDWELTROUNDABOUTTHEMANDALLT
HESESAYINGSWERENOISEDABROADTHROUGHOUTALLTHEHIL
LCOUNTRYOFJUDAEAANDALLTHATHEARDTHEMLAIDTHEMUPI
```

Verses in Matrix

The Gospel According to St. Luke 1:53-1:64

Lilies on the casket above: another flower with the Master 22 vibration:

L I L Y
3 9 3 7= 22 Diana's Master Number

The above matrix has no connection with the all-important Alma Tunnel/Pont and therefore, it could be claimed, the "Diana" in it could be anyone at all. However, in light of the preceding matrix which establishes that "Diana" is Diana Princess of Wales, then this matrix may be seen as supporting evidence/material. If the Bible Code is true, then we might assume Diana's final wishes were not granted and she was not buried in accordance with her specific request. In the matrix to follow, both the words "hospital" and the French equivalent "hopital" can be found equidistantly spaced in this matrix, incidentally, at different points.

Once again, the "Diana" above could be anybody at all, but connects well with the alleged evidence given to a medic at her side, concerning Princess Diana's final request. It is very possible of course, that those responsible for not passing on this message could have seen problems being posed with a

Muslim/Christian joint burial for Diana and Dodi. This would be especially so, given Diana's high profile, her connection with the royal family, and her role as mother of heirs to the throne. In light of the above considerations, in order to avoid huge complications, they may have decided it would be simpler just to ignore Diana's last wishes.

After all, they would have reasoned, the wishes of millions would have to be considered, and that would be that Diana should have a funeral befitting that of a royal Princess. Diana's surviving children would have to be considered above all, and no doubt they would have preferred to see their mother receive due recognition of her services to the world. That probably was best achieved by having the funeral that she did have.

However, her last resting place at her former home of Althorp seems to be one that she may not necessarily have desired. If she stated that she wanted to be with Dodi, and she was pregnant with his child, and this can be proven at some time forward, it behoves all of us who loved her, and the work she undertook on our behalf, (not just for herself), to see that her last wish is in some way, one day, granted. Now we move on to the least palatable of all the considerations.

Doctor Hasnat Khan of whom Diana's mother bitterly disapproved.

Section 29. Was any member of the royal family responsible for Diana's and Dodi's deaths?

Matches for: Diana dies Alma Tunnel fatherinlaw order car tip

Match: 1 of 1 Skip to Match

ELS (Equidistant Letter Spacing) Matrix

```
THSWORNJEHOVAHWILLHAVEWARWITHAMALEKFROMGENERAT
IONTOGENERATIONNOWJETHROTHEPRIESTOFMIDIANMOSES
FATHERINLAWHEARDOFALLTHATGODHADDONEFORMOSESAND
FORISRAELHISPEOPLEHOWTHATJEHOVAHHADBROUGHTISRA
ELOUTOFEGYPTANDJETHROMOSESFATHERINLAWTOOKZIPPO
RAHMOSESWIFEAFTERHEHADSENTHERAWAYANDHERTWOSONS
OFWHOMTHENAMEOFTHEONEWASGERSHOMFORHESAIDIHAVEB
EENASOJOURNERINAFOREIGNLANDANDTHENAMEOFTHEOTHE
RWASELIEZERFORHESAIDTHEGODOFMYFATHERWASMYHELPA
NDDELIVEREDMEFROMTHESWORDOFPHARAOHANDJETHROMOS
ESFATHERINLAWCAMEWITHHISSONSANDHISWIFEUNTOMOSE
SINTOTHEWILDERNESSWHEREHEWASENCAMPEDATTHEMOUNT
OFGODANDHESAIDUNTOMOSESITHYFATHERINLAWJETHROAM
COMEUNTOTHEEANDTHYWIFEANDHERTWOSONSWITHHERANDM
OSESWENTOUTTOMEETHISFATHERINLAWANDDIDOBEISANCE
ANDKISSEDHIMANDTHEYASKEDEACHOTHEROFTHEIRWELFAR
EANDTHEYCAMEINTOTHETENTANDMOSESTOLDHISFATHERIN
LAWALLTHATJEHOVAHHADDONEUNTOPHARAOHANDTOTHEEGY
PTIANSFORISRAELSSAKEALLTHETRAVAILTHATHADCOMEUP
ONTHEMBYTHEWAYANDHOWJEHOVAHDELIVEREDTHEMANDJET
```

Verses in Matrix

The Second Book of Moses, Called Exodus 17:16-18:8

Now, N.B.: it just so happens that we can also find the word "not" in a matrix, if we look at the verses 15:6 – 18:8, rendering false the statement that Diana's father in law ordered a car tip in the Alma Tunnel.

Matches for: Diana Alma Tunnel fatherinlaw not order car tip

Match: 1 of 1 Skip to Match

ELS (Equidistant Letter Spacing) Matrix

```
NDMOSESBUILTANALTARANDCALLEDTHENAMEOFITJEHOVAH
NISSIANDHESAIDJEHOVAHHATHSWORNJEHOVAHWILLHAVEW
ARWITHAMALEKFROMGENERATIONTOGENERATIONNOWJETHR
OTHEPRIESTOFMIDIANMOSESFATHERINLAWHEARDOFALLTH
ATGODHADDONEFORMOSESANDFORISRAELHISPEOPLEHOWTH
ATJEHOVAHHADBROUGHTISRAELOUTOFEGYPTANDJETHROMO
SESFATHERINLAWTOOKZIPPORAHMOSESWIFEAFTERHEHADS
ENTHERAWAYANDHERTWOSONSOFWHOMTHENAMEOFTHEONEWA
SGERSHOMFORHESAIDIHAVEBEENASOJOURNERINAFOREIGN
LANDANDTHENAMEOFTHEOTHERWASELIEZERFORHESAIDTHE
GODOFMYFATHERWASMYHELPANDDELIVEREDMEFROMTHESWO
RDOFPHARAOHANDJETHROMOSESFATHERINLAWCAMEWITHHI
SSONSANDHISWIFEUNTOMOSESINTOTHEWILDERNESSWHERE
HEWASENCAMPEDATTHEMOUNTOFGODANDHESAIDUNTOMOSES
ITHYFATHERINLAWJETHROAMCOMEUNTOTHEEANDTHYWIFEA
NDHERTWOSONSWITHHERANDMOSESWENTOUTTOMEETHISFAT
HERINLAWANDDIDOBEISANCEANDKISSEDHIMANDTHEYASKE
DEACHOTHEROFTHEIRWELFAREANDTHEYCAMEINTOTHETENT
ANDMOSESTOLDHISFATHERINLAWALLTHATJEHOVAHHADDON
```

Verses in Matrix

The Second Book of Moses, Called Exodus 17:15-18:8

The words "father-in-law" in the matrix above are in the ordinary text of The Bible, and are therefore open to question as a legitimate search tool. In the following matrices, we look at the charges laid at the feet of both Philip and HRH Queen Elizabeth II.

Quote: Al-Fayed funded Diana movie alleges Prince Philip is a 'Fred West-style psychopath' and labels Royal Family 'gangsters in tiaras'
By LIZ THOMAS UPDATED: 01:42 GMT, 14 May 2011

The Royal Family masterminded the car crash that caused Princess Diana's death as a 'warning' to make her toe the line, a controversial new film premiered at Cannes yesterday claims. The director, Keith Allen, said the House of Windsor had 'got away with murder' because no one had been held to account over the accident.

In the film, entitled Unlawful Killing, Prince Philip is branded a 'Fred West-style psychopath' who 'orchestrated the murder' of Diana. The Queen is dismissed as a 'gangster in a tiara'. End of Quote.

Reference: http://www.dailymail.co.uk/tvshowbiz/article-1386973/New-Diana-movie-calls-Prince-Phillip-Fred-West-style-psychopath-labels-Royal-Family-gangsters-tiaras.html#ixzz1s69DmJwS

In an interesting aside, the car hit the 13th pillar. The number 13 equates with The Death Card in the Tarot Deck which contains the Roman numerals XIII = 13 on the flag that the skeleton carries aloft. The sun rising in the distance reminds us that, though they pass from this earth, all things must die and be reborn anew on a heavenly plane.

It would of course be well nigh impossible to plan hitting any particular pillar in the tunnel, even if the crash had been engineered by intelligence agents. Yet there is a higher intelligence than our own who knows our fate, (and if the Bible Code is true), even at the hands of heinous criminals shielded from view among us, perhaps concealed in the guise of allies and friends.

Faces of Diana, Princess of Wales, as the world first knew her, a nineteen year old innocent bride. Put your hand over the right side of her face, then over the left. Her face is markedly asymmetrical. The eyes are windows to the soul. Her left eye shows her inner suffering, her right her compassion. The asymmetry is not so noticeable in this coquettish pose, right, taken after maturity and motherhood, which inevitably changes a young girl into a woman. The happiness and fulfilment motherhood brings is apparent in her face.

- ✓ Any sane person would have left long ago. But I cannot. I have my sons.
- ✓ Anywhere I see suffering, that is where I want to be, doing what I can.
- ✓ Being a princess isn't all it's cracked up to be.
- ✓ Carry out a random act of kindness, with no expectation of reward, safe in the knowledge that one day someone might do the same for you.
- ✓ Everyone of us needs to show how much we care for each other and, in the process, care for ourselves.
- ✓ Family is the most important thing in the world.
- ✓ HIV does not make people dangerous to know, so you can shake their hands and give them a hug: Heaven knows they need it.
- ✓ Hugs can do great amounts of good - especially for children.
- ✓ I don't even know how to use a parking meter, let alone a phone box.
- ✓ I don't go by the rule book... I lead from the heart, not the head.
- ✓ I don't want expensive gifts; I don't want to be bought. I have everything I want. I just want someone to be there for me, to make me feel safe and secure.
- ✓ I knew what my job was; it was to go out and meet the people and love them.
- ✓ I like to be a free spirit. Some don't like that, but that's the way I am.
- ✓ I live for my sons. I would be lost without them.
- ✓ I think like any marriage, especially when you've had divorced parents like myself; you want to try even harder to make it work.
- ✓ I think the biggest disease the world suffers from in this day and age is the disease of people feeling unloved. I know that I can give love for a minute, for half an hour, for a day, for a month, but I can give. I am very happy to do that, I want to do that.
- ✓ I want my boys to have an understanding of people's emotions, their insecurities, people's distress, and their hopes and dreams.
- ✓ I want to walk into a room, be it a hospital for the dying or a hospital for the sick children, and feel that I am needed. I want to do, not just to be.
- ✓ I wear my heart on my sleeve.
- ✓ I will fight for my children on any level so they can reach their potential as human beings and in their public duties.
- ✓ I'd like to be a queen in people's hearts but I don't see myself being queen of this country.
- ✓ I'm as thick as a plank.
- ✓ I'm aware that people I have loved and have died and are in the spirit world looking after me.
- ✓ If men had to have babies, they would only ever have one each.
- ✓ If you find someone you love in your life, then hang on to that love.
- ✓ It's vital that the monarchy keeps in touch with the people. It's what I try and do.
- ✓ Life is just a journey.

- ✓ Nothing brings me more happiness than trying to help the most vulnerable people in society. It is a goal and an essential part of my life - a kind of destiny. Whoever is in distress can call on me. I will come running wherever they are.
- ✓ Only do what your heart tells you.
- ✓ People think that at the end of the day a man is the only answer. Actually, a fulfilling job is better for me.
- ✓ So many people supported me through my public life and I will never forget them.
- ✓ The biggest disease this day and age is that of people feeling unloved.
- ✓ The greatest problem in the world today is intolerance. Everyone is so intolerant of each other.
- ✓ The kindness and affection from the public have carried me through some of the most difficult periods, and always your love and affection have eased the journey.
- ✓ There were three of us in this marriage, so it was a bit crowded.
- ✓ They say it is better to be poor and happy than rich and miserable, but how about a compromise like moderately rich and just moody?
- ✓ What must it be like for a little boy to read that daddy never loved mummy?
- ✓ When you are happy you can forgive a great deal.
- ✓ You can't comfort the afflicted with afflicting the comfortable.

Reference:
http://www.brainyquote.com/quotes/authors/p/princess_diana_2.html

Eulogy for Princess Diana by Her Brother Charles Westminster Abbey, London, September 6, 1997

I stand before you today the representative of a family in grief, in a country in mourning before a world in shock.

We are all united not only in our desire to pay our respects to Diana but rather in our need to do so.

For such was her extraordinary appeal that the tens of millions of people taking part in this service all over the world via television and radio who never actually met her, feel that they, too, lost someone close to them in the early hours of Sunday morning. It is a more remarkable tribute to Diana than I can ever hope to offer her today.

Diana was the very essence of compassion, of duty, of style, of beauty. All over the world she was a symbol of selfless humanity, a standard-bearer for the rights of the truly downtrodden, a very British girl who transcended nationality, someone with a natural nobility who was classless, who proved in the last year that she needed no royal title to continue to generate her particular brand of magic.

Today is our chance to say "thank you" for the way you brightened our lives, even though God granted you but half a life. We will all feel cheated, always, that you were taken from us so young and yet we must learn to be grateful that you came along at all.

Only now you are gone do we truly appreciate what we are now without and we want you to know that life without you is very, very difficult.

We have all despaired at our loss over the past week and only the strength of the message you gave us through your years of giving has afforded us the strength to move forward.

There is a temptation to rush to canonize your memory. There is no need to do so. You stand tall enough as a human being of unique qualities not to need to be seen as a saint. Indeed to sanctify your memory would be to miss out on the very core of your being, your wonderfully mischievous sense of humour with the laugh that bent you double, your joy for life transmitted wherever you took your smile, and the sparkle in those unforgettable eyes, your boundless energy which you could barely contain.

But your greatest gift was your intuition, and it was a gift you used wisely. This is what underpinned all your wonderful attributes. And if we look to analyze what it was about you that had such a wide appeal, we find it in your instinctive feel for what was really important in all our lives.

Without your God-given sensitivity, we would be immersed in greater ignorance at the anguish of AIDS and HIV sufferers, the plight of the homeless, the isolation of lepers, the random destruction of land mines. Diana explained to me once that it was her innermost feelings of suffering that made it possible for her to connect with her constituency of the rejected.

And here we come to another truth about her. For all the status, the glamour, the applause, Diana remained throughout a very insecure person at heart, almost childlike in her desire to do good for others so she could release herself from deep feelings of unworthiness of which her eating disorders were merely a symptom.

The world sensed this part of her character and cherished her for her vulnerability, whilst admiring her for her honesty. The last time I saw Diana was on July the first, her birthday, in London, when typically she was not taking time to celebrate her special day with friends but was guest of honour at a fund-raising charity evening.

She sparkled of course, but I would rather cherish the days I spent with her in March when she came to visit me and my children in our home in South Africa. I am proud of the fact that apart from when she was on public display meeting President Mandela, we managed to contrive to stop the ever-present paparazzi from getting a single picture of her. That meant a lot to her.

These were days I will always treasure. It was as if we'd been transported back to our childhood, when we spent such an enormous amount of time together, the two youngest in the family.

Fundamentally she hadn't changed at all from the big sister who mothered me as a baby, fought with me at school and endured those long train journeys between our parents' homes with me at weekends. It is a tribute to her level-headedness and strength that despite the most bizarre life imaginable after her childhood, she remained intact, true to herself.

There is no doubt that she was looking for a new direction in her life at this time. She talked endlessly of getting away from England, mainly because of the treatment she received at the hands of the newspapers.

I don't think she ever understood why her genuinely good intentions were sneered at by the media, why there appeared to be a permanent quest on their behalf to bring her down. It is baffling. My own, and only, explanation is that genuine goodness is threatening to those at the opposite end of the moral spectrum.

It is a point to remember that of all the ironies about Diana, perhaps the greatest was this; that a girl given the name of the ancient goddess of hunting was, in the end, the most hunted person of the modern age.

She would want us today to pledge ourselves to protecting her beloved boys William and Harry from a similar fate. And I do this here, Diana, on your behalf. We will not allow them to suffer the anguish that used regularly to drive you to tearful despair.

Beyond that, on behalf of your mother and sisters, I pledge that we, your blood family, will do all we can to continue the imaginative and loving way in which you were steering these two exceptional young men, so that their souls are not simply immersed by duty and tradition but can sing openly as you planned.

We fully respect the heritage into which they have both been born, and will always respect and encourage them in their royal role. But we, like you, recognize the need for them to experience as many different aspects of life as possible, to arm them spiritually and emotionally for the years ahead. I know you would have expected nothing less from us.

William and Harry, we all care desperately for you today. We are all chewed up with sadness at the loss of a woman who wasn't even our mother. How great your suffering is we cannot even imagine.

I would like to end by thanking God for the small mercies he has shown us at this dreadful time; for taking Diana at her most beautiful and radiant and when she had joy in her private life.

Above all, we give thanks for the life of a woman I am so proud to be able to call my sister: the unique the complex, the extraordinary and irreplaceable Diana, whose beauty, both internal and external, will never be extinguished from our minds.

The rowboat that takes visitors to the site of Diana's burial place, on the tiny island in the middle of the lake at Althorp.

Matches for: Queen Philip plan MOD SIS Fiat road death not true

Match: 1 of 1 Skip to Match

ELS (Equidistant Letter Spacing) Matrix

```
NTHEEFORISEETHATTHOUARTINTHEGALLOFBITTERNESSAN
DINTHEBONDOFINIQUITYANDSIMONANSWEREDANDSAIDPRA
YYEFORMETOTHELORDTHATNONEOFTHETHINGSWHICHYEHAV
ESPOKENCOMEUPONMETHEYTHEREFOREWHENTHEYHADTESTI
FIEDANDSPOKENTHEWORDOFTHELORDRETURNEDTOJERUSAL
EMANDPREACHEDTHEGOSPELTOMANYVILLAGESOFTHESAMAR
ITANSBUTANANGELOFTHELORDSPAKEUNTOPHILIPSAYINGA
RISEANDGOTOWARDTHESOUTHUNTOTHEWAYTHATGOETHDOWN
FROMJERUSALEMUNTOGAZATHESAMEISDESERTANDHEAROSE
ANDWENTANDBEHOLDAMANOFETHIOPIAAEUNUCHOFGREATAU
THORITYUNDERCANDACEQUEENOFTHEETHIOPIANSWHOWASO
VERALLHERTREASUREWHOHADCOMETOJERUSALEMTOWORSHI
PANDHEWASRETURNINGANDSITTINGINHISCHARIOTANDWAS
READINGTHEPROPHETISAIAHANDTHESPIRITSAIDUNTOPHI
LIPGONEARANDJOINTHYSELFTOTHISCHARIOTANDPHILIPR
ANTOHIMANDHEARDHIMREADINGISAIAHTHEPROPHETANDSA
IDUNDERSTANDESTTHOUWHATTHOUREADESTANDHESAIDHOW
CANIEXCEPTSOMEONESHALLGUIDEMEANDHEBESOUGHTPHIL
IPTOCOMEUPANDSITWITHHIMNOWTHEPASSAGEOFTHESCRIP
TUREWHICHHEWASREADINGWASTHISHEWASLEDASASHEEPTO
```

Verses in Matrix

The Acts of the Apostles 8:22-8:30

Although now shown here, the words "car" and "seat" also come up equidistantly in the above matrix, in lines 1, 2 (seat) and 11 (car). The words "Queen" and "Philip" come up in connection with the charges only by using the ordinary text of The Bible, rendering the results very much open to question in the minds of some.

However, the Matrix below holds no ordinary Biblical text within its results. Dodi's father claimed that investigations after the crash revealed that his driver Henri Paul had been an intelligence agent, used by his employers without his knowledge, in effect, to act as a suicide driver.

The two families, the Windsors and the Spencers, gather for the first time since Diana's death: for the opening of the Diana memorial in Hyde Park.

Memorial plaque at the site, showing the symbol of the rose. Diana, referred to as England's Rose by Elton John in his song of dedication, sung at her funeral. USA President Bill Clinton, with whom Diana met just prior to her death, to implore him to ban land mines. He went to the Oslo Conference days after she died and voted to retain them.

Matches for: not true SIS Henri Paul agent used die Alma false

Match: 1 of 24 Skip to Match

ELS (Equidistant Letter Spacing) Matrix

```
SAITHJEHOVAHTHECHILDRENOFISRAELSHALLCOMETHEYAN
DTHECHILDRENOFJUDAHTOGETHERTHEYSHALLGOONTHEIRW
AYWEEPINGANDSHALLSEEKJEHOVAHTHEIRGODTHEYSHALLI
NQUIRECONCERNINGZIONWITHTHEIRFACESTHITHERWARDS
AYINGCOMEYEANDJOINYOURSELVESTOJEHOVAHINANEVERL
ASTINGCOVENANTTHATSHALLNOTBEFORGOTTENMYPEOPLEH
AVEBEENLOSTSHEEPTHEIRSHEPHERDSHAVECAUSEDTHEMTO
GOASTRAYTHEYHAVETURNEDTHEMAWAYONTHEMOUNTAINSTH
EYHAVEGONEFROMMOUNTAINTOHILLTHEYHAVEFORGOTTENT
HEIRRESTINGPLACEALLTHATFOUNDTHEMHAVEDEVOUREDTH
EMANDTHEIRADVERSARIESSAIDWEARENOTGUILTYBECAUSE
THEYHAVESINNEDAGAINSTJEHOVAHTHEHABITATIONOFRIG
HTEOUSNESSEVENJEHOVAHTHEHOPEOFTHEIRFATHERSFLEE
OUTOFTHEMIDSTOFBABYLONANDGOFORTHOUTOFTHELANDOF
THECHALDEANSANDBEASTHEHEGOATSBEFORETHEFLOCKSFO
RLOIWILLSTIRUPANDCAUSETOCOMEUPAGAINSTBABYLONAC
OMPANYOFGREATNATIONSFROMTHENORTHCOUNTRYANDTHEY
SHALLSETTHEMSELVESINARRAYAGAINSTHERFROMTHENCES
HESHALLBETAKENTHEIRARROWSSHALLBEASOFANEXPERTMI
GHTYMANNONESHALLRETURNINVAINANDCHALDEASHALLBEA
PREYALLTHATPREYUPONHERSHALLBESATISFIEDSAITHJEH
OVAHBECAUSEYEAREGLADBECAUSEYEREJOICEOYETHATPLU
NDERMYHERITAGEBECAUSEYEAREWANTONASAHEIFERTHATT
READETHOUTTHEGRAINANDNEIGHASSTRONGHORSESYOURMO
THERSHALLBEUTTERLYPUTTOSHAMESHETHATBAREYOUSHAL
LBECONFOUNDEDBEHOLDSHESHALLBETHEHINDERMOSTOFTH
ENATIONSAWILDERNESSADRYLANDANDADESERTBECAUSEOF
```

Verses in Matrix

The Book of the Prophet Jeremiah 50:4-50:12

The Diana, Princess of Wales Memorial Fountain (above) in Hyde Park London, designed by American architect Kathryn Gustafson, cost £3.6m and is based on an 80 metre x 50 metre oval granite stone ring, around which water flows in “a never-ending river of life”.

As you can see, there are 24 matches in The Bible for this matrix. The word “Merc” also comes up equidistantly in the one of the matrices to be found, in the last four lines of Leviticu 20:6. If The Bible Code is true, the double negation of “not true” and “false” above renders the assertion “Henri Paul was a SIS agent that died in the Alma in a Merc” necessarily true.

Matches for: hidden MOD SIS Diana death Alma Pont not true

Match: 1 of 1 Skip to Match

ELS (Equidistant Letter Spacing) Matrix

```
NTOTHECHIEFSOFTHEMENOFWARTHATWENTWITHHIMCOMENE
ARPUTYOURFEETUPONTHENECKSOFTHESEKINGSANDTHEYCA
MENEARANDPUTTHEIRFEETUPONTHENECKSOFTHEMANDJOSH
UASAIDUNTOTHEMFEARNOTNORBEDISMAYEDBESTRONGANDO
FGOODCOURAGEFORTHUSSHALLJEHOVAHDOTOALLYOURENEM
IESAGAINSTWHOMYEFIGHTANDAFTERWARDJOSHUASMOTETH
EMANDPUTTHEMTODEATHANDHANGEDTHEMONFIVETREESAND
THEYWEREHANGINGUPONTHETREESUNTILTHEEVENINGANDI
TCAMETOPASSATTHETIMEOFTHEGOINGDOWNOFTHESUNTHAT
JOSHUACOMMANDEDANDTHEYTOOKTHEMDOWNOFFTHETREESA
NDCASTTHEMINTOTHECAVEWHEREINTHEYHADHIDDENTHEMS
ELVESANDLAIDGREATSTONESONTHEMOUTHOFTHECAVEUNTO
THISVERYDAYANDJOSHUATOOKMAKKEDAHONTHATDAYANDSM
OTEITWITHTHEEDGEOFTHESWORDANDTHEKINGTHEREOFHEU
TTERLYDESTROYEDTHEMANDALLTHESOULSTHATWERETHERE
INHELEFTNONEREMAININGANDHEDIDTOTHEKINGOFMAKKED
AHASHEHADDONEUNTOTHEKINGOFJERICHOANDJOSHUAPASS
EDFROMMAKKEDAHANDALLISRAELWITHHIMUNTOLIBNAHAND
FOUGHTAGAINSTLIBNAHANDJEHOVAHDELIVEREDITALSOAN
DTHEKINGTHEREOFINTOTHEHANDOFISRAELANDHESMOTEIT
WITHTHEEDGEOFTHESWORDANDALLTHESOULSTHATWERETHE
REINHELEFTNONEREMAININGINITANDHEDIDUNTOTHEKING
THEREOFASHEHADDONEUNTOTHEKINGOFJERICHOANDJOSHU
APASSEDFROMLIBNAHANDALLISRAELWITHHIMUNTOLACHIS
```

Verses in Matrix

The Book of Joshua 10:24-10:31

The words "hidden" and "death" are in the ordinary text and therefore, according to some researchers, not legitimately connected with this search. The word "death" crosses "true", giving weight according to some. Photo below: cleaning up cluster mines, left behind after war has ended. Diana wasn't just a pretty face, was she now?

Water flows quietly along the Diana Memorial Fountain on an overcast, rainy day in Hyde Park London. The Roman Goddess Diana.

Matches for: MOD SIS cause Diana's death Alma Pont not true

Match: 1 of 12 Skip to Match

ELS (Equidistant Letter Spacing) Matrix

```
NDEDANDITSTOODFASTJEHOVAHBRINGETHTHECOUNSELOFT
HENATIONSTONOUGHTHEMAKETHTHETHOUGHTSOFTHEPEOPL
ESTOBEOFNOEFFECTTHECOUNSELOFJEHOVAHSTANDETHFAS
TFOREVERTHETHOUGHTSOFHISHEARTTOALLGENERATIONSB
LESSEDISTHENATIONWHOSEGODISJEHOVAHTHEPEOPLEWHO
MHEHATHCHOSENFORHISOWNINHERITANCEJEHOVAHLOOKET
HFROMHEAVENHEBEHOLDETHALLTHESONSOFMENFROMTHEPL
ACEOFHISHABITATIONHELOOKETHFORTHUPONALLTHEINHA
BITANTSOFTHEEARTHHETHATFASHIONETHTHEHEARTSOFTH
EMALLTHATCONSIDERETHALLTHEIRWORKSTHEREISNOKING
SAVEDBYTHEMULTITUDEOFAHOSTAMIGHTYMANISNOTDELIV
EREDBYGREATSTRENGTHAHORSEISAVAINTHINGFORSAFETY
NEITHERDOTHHEDELIVERANYBYHISGREATPOWERBEHOLDTH
EEYEOFJEHOVAHISUPONTHEMTHATFEARHIMUPONTHEMTHAT
HOPEINHISLOVINGKINDNESSTODELIVERTHEIRSOULFROMD
EATHANDTOKEEPTHEMALIVEINFAMINEOURSOULHATHWAITE
DFORJEHOVAHHEISOURHELPANDOURSHIELDFOROURHEARTS
HALLREJOICEINHIMBECAUSEWEHAVETRUSTEDINHISHOLYN
AMELETTHYLOVINGKINDNESSOJEHOVAHBEUPONUSACCORDI
NGASWEHAVEHOPEDINTHEEPSALMAPSALMOFDAVIDWHENHEC
HANGEDHISBEHAVIORBEFOREABIMELECHWHODROVEHIMAWA
YANDHEDEPARTEDIWILLBLESSJEHOVAHATALLTIMESHISPR
AISESHALLCONTINUALLYBEINMYMOUTHMYSOULSHALLMAKE
```

Verses in Matrix

The Book of Psalms Book I Psalm 1 33:9-34:1

The word "death" appears in the ordinary Biblical text, but it does not intersect or cross any other text. I think given the above, we must logically draw the conclusion that, if The Bible Code is true, then neither SIS nor MOD is responsible for Diana's death, as the word "false" cannot also be added to this matrix.

I have mentioned before the significance for me in the numbering (33, 34) of the verses above. In this case, the ordinary Biblical text seems very potent. It refers to the conspiracy against King Amon, resulting in his murder. "And his servants conspired against him, and put him to death in his own house. But the people of the land slew all them that had conspired against King Amon and the people of the land made Josiah his son King in his stead."

King Amon of Judah from the Royal House of David – c664-c641BC

Baby bump, pre-menstrual tummy or middle aged spread? Diana in one of the last pictures taken of her - arguably in July 1997 - happy holidaying with Dodi, sporting a snazzy animal print swimsuit, every inch the "Yummy Mummy". Just as the Roman Goddess Diana (Artemis is the equivalent in Greek mythology) carried a hunter's bow and wore leopard skin, so did our Princess Diana.

Artemis/Diana was the great Olympian goddess of hunting, wilderness and wild animals. She was also a goddess of childbirth, and the protectress of the girl child up to the age of marriage. Her twin brother Apollo was similarly the protector of the boy child. Together the two gods were also bringers of sudden death and disease - Artemis targeted women and girls, Apollo men and boys.

Section 30. The USA, Diana, and the landmines issue.

Matches for: USA CIA FBI Diana land mine ban not caused death

Match: 1 of 24 Skip to Match

ELS (Equidistant Letter Spacing) Matrix

```
NTODRESSITANDTOKEEPITANDJEHOVAHGODCOMMANDEDTHE
MANSAYINGOFEVERYTREEOFTHEGARDENTHOUMAYESTFREEL
YEATBUTOFTHETREEOFTHEKNOWLEDGEOFGOODANDEVILTHO
USHALTNOTEATOFITFORINTHEDAYTHATTHOUEATESTTHERE
OFTHOUSHALTSURELYDIEANDJEHOVAHGODSAIDITISNOTGO
ODTHATTHEMANSHOULDBEALONEIWILLMAKEHIMAHELPMEET
FORHIMANDOUTOFTHEGROUNDJEHOVAHGODFORMEDEVERYBE
ASTOFTHEFIELDANDEVERYBIRDOFTHEHEAVENSANDBROUGH
TTHEMUNTOTHEMANTOSEEWHATHEWOULDCALLTHEMANDWHAT
SOEVERTHEMANCALLEDEVERYLIVINGCREATURETHATWASTH
ENAMETHEREOFANDTHEMANGAVENAMESTOALLCATTLEANDTO
THEBIRDSOFTHEHEAVENSANDTOEVERYBEASTOFTHEFIELDB
UTFORMANTHEREWASNOTFOUNDAHELPMEETFORHIMANDJEHO
VAHGODCAUSEDADEEPSLEEPTOFALLUPONTHEMANANDHESLE
PTANDHETOOKONEOFHISRIBSANDCLOSEDUPTHEFLESHINST
EADTHEREOFANDTHERIBWHICHJEHOVAHGODHADTAKENFROM
THEMANMADEHEAWOMANANDBROUGHTHERUNTOTHEMANANDTH
EMANSAIDTHISISNOWBONEOFMYBONESANDFLESHOFMYFLES
HSHESHALLBECALLEDWOMANBECAUSESHEWASTAKENOUTOFM
ANTHEREFORESHALLAMANLEAVEHISFATHERANDHISMOTHER
ANDSHALLCLEAVEUNTOHISWIFEANDTHEYSHALLBEONEFLES
HANDTHEYWEREBOTHNAKEDTHEMANANDHISWIFEANDWERENO
TASHAMEDNOWTHESERPENTWASMORESUBTLETHANANYBEAST
OFTHEFIELDWHICHJEHOVAHGODHADMADEANDHESAIDUNTOT
HEWOMANYEAHATHGODSAIDYESHALLNOTEATOFANYTREEOFT
HEGARDENANDTHEWOMANSAIDUNTOTHESERPENTOFTHEFRUI
TOFTHETREESOFTHEGARDENWEMAYEATBUTOFTHEFRUITOFT
HETREEWHICHISINTHEMIDSTOFTHEGARDENGODHATHSAIDY
```

Verses in Matrix

The First Book of Moses, called Genesis 2:15-3:3

Althorp

Matches for: CIA FBI kill Diana over land mines ban not true

Match: 1 of 8 Skip to Match

ELS (Equidistant Letter Spacing) Matrix

```
BYTHELOINSANDTHECAULUPONTHELIVERWITHTHEKIDNEYS
SHALLHETAKEAWAYANDTHEPRIESTSHALLBURNITUPONTHEA
LTARITISTHEFOODOFTHEOFFERINGMADEBYFIREUNTOJEHO
VAHANDIFHISOBLATIONBEAGOATTHENHESHALLOFFERITBE
FOREJEHOVAHANDHESHALLLAYHISHANDUPONTHEHEADOFIT
ANDKILLITBEFORETHETENTOFMEETINGANDTHESONSOFAAR
ONSHALLSPRINKLETHEBLOODTHEREOFUPONTHEALTARROUN
DABOUTANDHESHALLOFFERTHEREOFHISOBLATIONEVENANO
FFERINGMADEBYFIREUNTOJEHOVAHTHEFATTHATCOVERETH
THEINWARDSANDALLTHEFATTHATISUPONTHEINWARDSANDT
HETWOKIDNEYSANDTHEFATTHATISUPONTHEMWHICHISBYTH
ELOINSANDTHECAULUPONTHELIVERWITHTHEKIDNEYSSHAL
LHETAKEAWAYANDTHEPRIESTSHALLBURNTHEMUPONTHEALT
ARITISTHEFOODOFTHEOFFERINGMADEBYFIREFORASWEETS
AVORALLTHEFATISJEHOVAHSITSHALLBEAPERPETUALSTAT
UTETHROUGHOUTYOURGENERATIONSINALLYOURDWELLINGS
THATYESHALLEATNEITHERFATNORBLOODANDJEHOVAHSPAK
EUNTOMOSESSAYINGSPEAKUNTOTHECHILDRENOFISRAELSA
YINGIFANYONESHALLSINUNWITTINGLYINANYOFTHETHING
SWHICHJEHOVAHHATHCOMMANDEDNOTTOBEDONEANDSHALLD
OANYONEOFTHEMIFTHEANOINTEDPRIESTSHALLSINSOASTO
```

Verses in Matrix

The Third Book of Moses, commonly called Leviticu 3:10-4:3

On the face of it, if The Bible Code is true, it would seem from the above that the USA, CIA and FBI played no part in Diana's death. However, look at the matrix below. When a statement is "not false", then it is necessarily true. Therefore, the assertion "The USA, CIA and FBI played a part in Diana's death because of her involvement in the land mines issue" becomes a necessarily true one.

Matches for: USA CIA FBI Diana land mine death caused not false

Match: 1 of 10 Skip to Match

ELS (Equidistant Letter Spacing) Matrix

```
UTIFTHEREISNORESURRECTIONOFTHEDEADNEITHERHATHC
HRISTBEENRAISEDANDIFCHRISTHATHNOTBEENRAISEDTHE
NISOURPREACHINGVAINYOURFAITHALSOISVAINYEAWEARE
FOUNDFALSEWITNESSESOFGODBECAUSEWEWITNESSEDOFGO
DTHATHERAISEDUPCHRISTWHOMHERAISEDNOTUPIFSOBETH
ATTHEDEADARENOTRAISEDFORIFTHEDEADARENOTRAISEDN
EITHERHATHCHRISTBEENRAISEDANDIFCHRISTHATHNOTBE
ENRAISEDYOURFAITHISVAINYEAREYETINYOURSINSTHENT
HEYALSOTHATAREFALLENASLEEPINCHRISTHAVEPERISHED
IFWEHAVEONLYHOPEDINCHRISTINTHISLIFEWEAREOFALLM
ENMOSTPITIABLEBUTNOWHATHCHRISTBEENRAISEDFROMTH
EDEADTHEFIRSTFRUITSOFTHEMTHATAREASLEEPFORSINCE
BYMANCAMEDEATHBYMANCAMEALSOTHERESURRECTIONOFTH
EDEADFORASINADAMALLDIESOALSOINCHRISTSHALLALLBE
MADEALIVEBUTEACHINHISOWNORDERCHRISTTHEFIRSTFRU
ITSTHENTHEYTHATARECHRISTSATHISCOMINGTHENCOMETH
THEENDWHENHESHALLDELIVERUPTHEKINGDOMTOGODEVENT
HEFATHERWHENHESHALLHAVEABOLISHEDALLRULEANDALLA
UTHORITYANDPOWERFORHEMUSTREIGNTILLHEHATHPUTALL
HISENEMIESUNDERHISFEETTHELASTENEMYTHATSHALLBEA
BOLISHEDISDEATHFORHEPUTALLTHINGSINSUBJECTIONUN
DERHISFEETBUTWHENHESAITHALLTHINGSAREPUTINSUBJE
```

Verses in Matrix

The First Epistle of Paul the Apostle to the Cori 15:13-15:27

Matches for: CIA FBI USA Diana death Alma Tunnel land mine ban

Match: 1 **of 16** Skip to Match

ELS (Equidistant Letter Spacing) Matrix

```
BECOMEOFHISDREAMSANDREUBENHEARDITANDDELIVEREDH
IMOUTOFTHEIRHANDANDSAIDLETUSNOTTAKEHISLIFEANDR
EUBENSAIDUNTOTHEMSHEDNOBLOODCASTHIMINTOTHISPIT
THATISINTHEWILDERNESSBUTLAYNOHANDUPONHIMTHATHE
MIGHTDELIVERHIMOUTOFTHEIRHANDTORESTOREHIMTOHIS
FATHERANDITCAMETOPASSWHENJOSEPHWASCOMEUNTOHISB
RETHRENTHATTHEYSTRIPPEDJOSEPHOFHISCOATTHECOATO
FMANYCOLORSTHATWASONHIMANDTHEYTOOKHIMANDCASTHI
MINTOTHEPITANDTHEPITWASEMPTYTHEREWASNOWATERINI
TANDTHEYSATDOWNTOEATBREADANDTHEYLIFTEDUPTHEIRE
YESANDLOOKEDANDBEHOLDACARAVANOFISHMAELITESWASC
OMINGFROMGILEADWITHTHEIRCAMELSBEARINGSPICERYAN
DBALMANDMYRRHGOINGTOCARRYITDOWNTOEGYPTANDJUDAH
SAIDUNTOHISBRETHRENWHATPROFITISITIFWESLAYOURBR
OTHERANDCONCEALHISBLOODCOMEANDLETUSSELLHIMTOTH
EISHMAELITESANDLETNOTOURHANDBEUPONHIMFORHEISOU
RBROTHEROURFLESHANDHISBRETHRENHEARKENEDUNTOHIM
ANDTHEREPASSEDBYMIDIANITESMERCHANTMENANDTHEYDR
EWANDLIFTEDUPJOSEPHOUTOFTHEPITANDSOLDJOSEPHTOT
HEISHMAELITESFORTWENTYPIECESOFSILVERANDTHEYBRO
UGHTJOSEPHINTOEGYPTANDREUBENRETURNEDUNTOTHEPIT
ANDBEHOLDJOSEPHWASNOTINTHEPITANDHERENTHISCLOTH
ESANDHERETURNEDUNTOHISBRETHRENANDSAIDTHECHILDI
```

Verses in Matrix

The First Book of Moses, called Genesis 37:20-37:30

CIA insignia left, FBI insignia on the right.

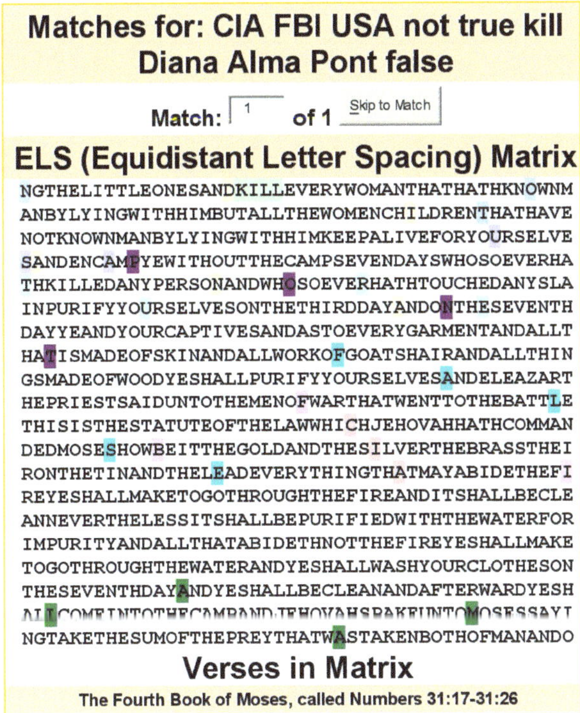

Quote: The US military may have played a role in the 1997 death of Princess Diana as she was campaigning for a worldwide ban on landmines, says the author of The Murder of Princess Diana. Noel Botham told Press TV that Princess Diana, during her visit to the US in 1997, succeeded in persuading former US President Bill Clinton to vote in favour of a global ban on landmines at the Oslo Conference, which was due to take place on the 19th of September of that year.

Reference: http://www.presstv.ir/detail/181945.html

However, the idea did not appeal to US military officials and they knew that Clinton "would not change his mind with Diana alive," said the author. "There was enormous pressure from the military in America to get him (President Clinton) to stop that (the international treaty on a worldwide ban on landmines) and change his mind, enormous lobbying in the White House," Botham added. "Of course, after her death, 19 days later, Bill Clinton went to Oslo and voted against a ban on landmines," he further explained.

The author of The Murder of Princess Diana also noted that in 1998, less than a year after Diana's death, news media around the world tried to get the CIA, the FBI and the National Security Agency to release their papers on Diana's death, which was to no avail. "They (the CIA, the FBI and the National Security Agency) all came up with one answer; that they would keep back certain pages, 145 pages that would seriously jeopardize the security of the US," Botham went on to say. Meanwhile, a movie called Unlawful Killing caused quite a stir in this year's Cannes Film Festival in France.

It is the story of the deaths of Princess Diana, Dodi Fayed and their driver Henri Paul, and was described by its director Keith Allen as "inquest into the inquest of Princess Diana." The film cannot be screened in the UK over 'national security issues.' End of Quote.

Left: Keith Allen Filmmaker Right: Noel Botham Author

Quote: Unlawful Killing – the film the British won't get to see. My documentary about the Diana inquest will be shown everywhere but the UK. Here's why.

The internet is a global lavatory wall, a Rabelaisian mixture of truth, lies, insanity and humour. I felt its power and madness this week, when an excerpt from my new film, Unlawful Killing, was leaked on to YouTube and seized on by US conspiracy theorists, who immediately began claiming that the CIA had murdered Princess Diana, thereby allowing others to dismiss my documentary as mad.

Reference: http://www.guardian.co.uk/commentisfree/2011/may/07/unlawful-killing-film-you-wont-see

Deriding its critics as mad is an age-old British establishment trick. My "inquest of the inquest" film contains footage of Diana recalling how the royals wanted her consigned to a mental institution, and the inquest coroner repeatedly questioning the sanity of anyone who wondered if the crash was more than an accident. His chief target was Mohamed Al Fayed, a man I once profiled for a Channel 4 documentary.

Before I met him, I'd half-believed the media caricature of him as a madman, driven nuts by the death of his son, and wildly accusing the Windsors of having planned the 1997 crash. However, I found a man who was sane and funny but frustrated that Britain wouldn't hold an inquest into his son's death. Michael Mansfield QC thought it unfair too, and fought for one to be held; which was why the longest inquest in British legal history eventually began in 2007.

Long before the inquest started, the eminently sane Mansfield had persuaded me that there were suspicious circumstances surrounding the crash, and signs of a cover-up by the authorities. Many journalists agreed, but as the inquest drew near, I noticed that British newspapers (several of which had regularly run "Was Diana Murdered?" pieces) suddenly fell into line, and started insisting that the inquest was a waste of time.

They raised no protest when virtually all the key French witnesses refused to participate, nor did they find it odd that not one senior royal was ordered to appear, even though Diana had stated in a lawyer's note that the Windsors were planning an "accident" to her car. Nor did they raise the issue of possible bias when legal proceedings involving the integrity of the royal family were to be heard in the royal courts of justice before a coroner who'd sworn an oath of allegiance to the Queen.

I felt the need to raise it, so I asked every major UK broadcaster (BBC, ITV, C4, Five, Sky) to commission a TV documentary about the inquest. But they refused even to contemplate such a suggestion, so Associated Rediffusion and I began filming and financing it ourselves. Shortly before the inquest began, Fayed offered to fund our project, so we could make a feature-length cinema documentary instead. We agreed, on condition that we would report events in the way we saw them, and the deal was struck.

Unlawful Killing is not about a conspiracy before the crash, but a provable conspiracy after the crash. A conspiracy organized not by a single scheming arch-fiend, but collectively by the British establishment – judges, lawyers, politicians, police chiefs, secret services, even newspaper editors – all of whom have been appointed to their positions because they are "a safe pair of hands".

Just as compass needles all point north without being told to, so these people instinctively know what is expected of them when the state's interests are under threat and they act accordingly, quietly suppressing uncomfortable evidence or undermining the credibility of witnesses whose evidence contradicts the official narrative.

Consider just a fraction of what transpired. Over 100 significant witnesses were not called to the inquest, or refused to appear. Blood tests allegedly proving the drunkenness of the driver Henri Paul were deemed "biologically inexplicable" by a toxicologist. A British crash expert found that Diana's seat belt had not been working. And so on.

Strangest of all was the media coverage of the verdict. Inquest evidence showed conclusively that the crash was caused by an unidentified white Fiat Uno and several unidentified motorcycles, vehicles that were certainly not paparazzi, because uncontested police evidence confirmed that the paparazzi were nowhere near the tunnel at the time of the crash.

The jury understood this, bringing in a verdict of "unlawful killing" by unidentified "following vehicles"; yet within seconds, the BBC was misreporting that the jury had blamed the paparazzi, and the rest of the media meekly followed suit. Which is why – three years on – barely anyone realizes what the jury's troubling verdict really was.

Why is the film being premiered next week at Cannes, three years after the inquest ended? Because British lawyers insisted on 87 cuts before any UK release could be contemplated. So rather than butcher the film, or risk legal action, we're showing it in France, then the US, and everywhere except the UK. Pity, because at a time when the mindless sugar rush of the royal wedding has been sending British Republicans into a diabetic coma, it could act as a welcome antidote. End of Quote.

The Happy Couple William and Kate, at the announcement of their engagement.

Section 31. The Numerology Chart of Diana.

Below is the numerology chart for Diana, born Diana Frances Spencer on 1st July, 1961. The number ruling her life between the ages of birth and 27 is found by adding the full value of the first three letters of her name, to the month she was born. DIA+7 = 4+9+1+7 = 21/3. This corresponds to The World card in the Tarot Deck. The female figure is dancing. Diana had great talent as a dancer. The number three, to which the 21 reduces, rules birth, and as the mother of a future King, her role was a most important one.

The Master 22/4 (Symbolized by The Fool card in the Tarot Deck – this card is interchangeable with Zero) ruling her life from the age of 27 to 54, was the one under which she died. We find this number by adding the second three letters of her name to the day on which she was born. NAF+1 = 14+1+6+1 = Master 22/4. This number reveals both her talents and possibilities/potentialities. Master Numbers (multiples of 11 (eleven), and any number that is reducible to 11), are extremely powerful, and offer to their possessor master accomplishments.

They may even allow the possessor to rise to a position of world prominence through their work. A Master Number appearing in the placement that Diana has it, ruling her life between the ages of 27 and 54, definitely indicates this, and the kind of number appearing there will give some indication of the sphere in which she would rise. Diana was born into aristocracy, yet, relatively speaking, her role in the royal family advanced her onto the world stage. Prior to her death, she was reportedly the most photographed woman in the world.

I do not doubt Diana had decided natural ability as an actor, and I am sure this would be confirmed today by speaking to those who were closest to her. She had yet to explore her film star potential. Perhaps Costner picked up on this, which was why he offered her a starring role. The Master 22 has a particular link with the film world, and the actors within it. Usually an acting ability is pronounced.

F I L M
6 9 3 4 = Master 22/4

A C T O R S
1 3 2 6 9 1 = Master 22/4

(Each letter takes its value from its place in the alphabet. For example, the letter "L" as the 12th letter, has the value 12 = 1+2 =3.)

Dodi Fayed was a film producer: the executive producer of the films Chariots of Fire, Breaking Glass, F/X, F/X2, Hook, and The Scarlet Letter. He also served as the Executive Creative Consultant for the F/X television series. It seems certain Dodi had introduced Diana to a world where she could express herself in this domain.

Diana's divorce gave her freedom to, for the first time in her life, really project herself artistically. There is no doubt in my mind that the report given by Kevin Costner is true and correct. He was going to star Diana in the sequel to The Bodyguard.

Under The Fool card, Diana would have continued down a dangerous path, even if she fully comprehended the peril surrounding her (and it is quite likely that she did not clearly see the full extent of all directions from which attack could come). High ethics and ideals are a feature of this number. The inability to understand other's attitudes and ideas is also a problem for the Master 22 agent. Diana may have considered them backward, old-world, and become impatient with them. The truth of the matter was, she was ahead of her time, other-worldly, and to many she was unfathomable and mysterious. To the public, this would have generated adoration, but from her enemies, hatred and even a burning desire for revenge, particularly if she exposed them, as she was obviously set to do.

The Fool card has a reputation for outsmarting its critics: it is the genius in disguise as the fool: a genius for thinking in broad, sweeping terms, seeing what the public wants and needs, and knows exactly how to supply it.

To some onlookers, she would have appeared foolish, pursuing the course she blatantly did, and in particular, in giving up all that she could have had as the wife of Prince Charles, heir to the throne. A safer course to be sure. I am sure she asked herself the same question from time to time: "Why on earth am I doing this? Why have I made life so hard for myself?" It would have been difficult, maybe even especially so, for one born into a life of privilege. After all, she had a lot to lose, a lot to give up. Yet, she would have been driven on, knowing innately that she had a higher purpose, a higher quest to pursue and most importantly, a destiny to fulfil.

Charles, Diana's brother.

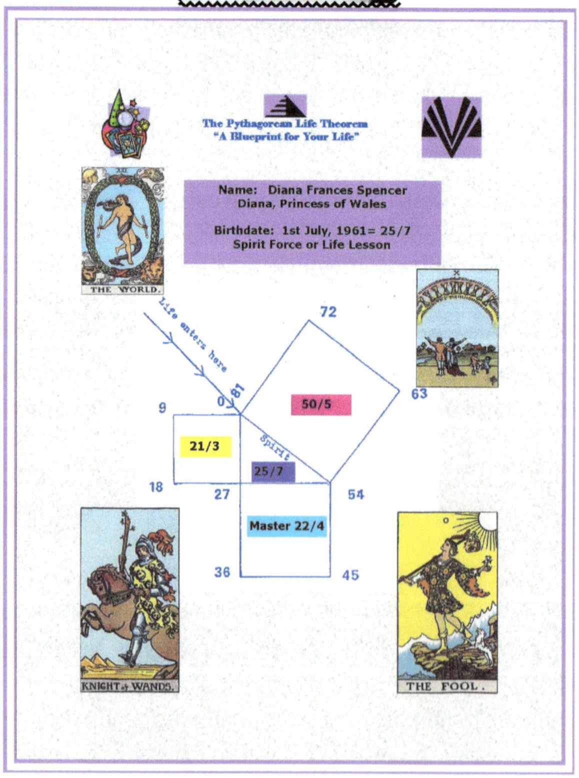

Note the pyramids in the Tarot Symbolism for the number 25, symbolised by the Knight of Wands. Diana was allegedly about to marry an Egyptian. This would be entirely fitting under this Life Lesson/Spirit Force vibration, although it is not the case that everyone with this card marries an Egyptian. As Dr. Hasnat Khan was a heart surgeon, it seems that Diana's Life Lesson, what she came here to learn in this lifetime, was directly connected with him. Certainly she would have had to learn that he was not up to the task of taking on her high profile, and instead wanted her to remain in the shadows with him. This would have sounded a death knell for their relationship, and eventually Diana realized this. I do not believe she was using Dodi as a tool to make Hasnat Khan sit up and take notice. No, she had moved on. She had found new love. I believe her destiny on this earth was ultimately connected with Dodi, an Egyptian by birth, as depicted in the card. She also may have had a natural propensity to befriend Muslims.

M U S L I M S
4 3 1 3 9 4 1 = 25/7 (Value of vowels = 12/3, Value of consonants=13/4)

Diana shared the number 25/7 with Marilyn Monroe, born on 1st June, 1926 = 1+6+(1+9+2+6=18) = 25/7. The word "heart" carries a 25/7 vibration:

Marilyn also had the number 25/7 ruling her life from age 27 to 54, and consequently, she died under this number. Both women died, by either injury to the heart, or heart failure. In Marilyn's case she reportedly received an injection directly into the heart, in an attempt to save her. Both were the same age at death - 36. Although Elton John may not have known the numerological connection of the 25/7, he certainly noticed other similarities, like their being the same age, and so adapted his song "Candle in the Wind" for Diana's funeral, the song he had written as a tribute to Marilyn Monroe.

It seems strangely poignant that both the original star of The Bodyguard, the beautiful, divinely talented singer Whitney Houston, and the equally beautiful Princess Diana, who would have taken the starring role in Bodyguard II, have died tragic deaths. Diana was a talented dancer, and even at the age of 36, it would not have been outside her capability to play a role that featured her dancing. I am sure she longed to express herself artistically, and this outlet had been long denied her in her role as future Queen of England and mother of heirs to the throne.

Much was made of Diana dancing on stage as a surprise birthday present for her husband Charles. The master copy of the performance is thought to be the only one in existence, rumoured to be owned by Charles himself. This was the kind of restriction placed on her artistic expression. The public was simply not allowed to see her perform, not while she was married to Charles. Not even now. The backlash at suggestions that she was to star in the Costner film and the rush to label him a liar for daring to even suggest it, shows that this suppression of her as a filmic or stage performer has continued even after her death.

Even her loyal butler, Paul Burrell, was askance at her "black legs" when she appeared, so un-English looking, un-Royal, deeply tanned, wearing white shorts, barefooted, *carrying* her sneakers, for God's sake, leaving the yacht with Dodi. The traditional, staid pale-skinned English image of the British monarchy, its adherence to strict protocol, must be preserved at all costs, and once again, even her pursuit of a cinematic career after her divorce from Charles could have been a very dangerous path for her to attempt to go down.

Matches for: Cost ner Body guard Two Diana star die too soon

Match: 1 **of 3** Skip to Match

ELS (Equidistant Letter Spacing) Matrix

```
GUARDCAMEANDBARETHEMANDBROUGHTTHEMBACKINTOTHEG
UARDCHAMBERANDWHENHEHUMBLEDHIMSELFTHEWRATHOFJE
HOVAHTURNEDFROMHIMSOASNOTTODESTROYHIMALTOGETHE
RANDMOREOVERINJUDAHTHEREWEREGOODTHINGSFOUNDSOK
INGREHOBOAMSTRENGTHENEDHIMSELFINJERUSALEMANDRE
IGNEDFORREHOBOAMWASFORTYANDONEYEARSOLDWHENHEBE
GANTOREIGNANDHEREIGNEDSEVENTEENYEARSINJERUSALE
MTHECITYWHICHJEHOVAHHADCHOSENOUTOFALLTHETRIBES
OFISRAELTOPUTHISNAMETHEREANDHISMOTHERSNAMEWASN
AAMAHTHEAMMONITESSANDHEDIDTHATWHICHWASEVILBECA
USEHESETNOTHISHEARTTOSEEKJEHOVAHNOWTHEACTSOFRE
HOBOAMFIRSTANDLASTARETHEYNOTWRITTENINTHEHISTOR
IESOFSHEMAIAHTHEPROPHETANDOFIDDOTHESEERAFTERTH
EMANNEROFGENEALOGIESANDTHEREWEREWARSBETWEENREH
OBOAMANDJEROBOAMCONTINUALLYANDREHOBOAMSLEPTWIT
HHISFATHERSANDWASBURIEDINTHECITYOFDAVIDANDABIJ
AHHISSONREIGNEDINHISSTEADINTHEEIGHTEENTHYEAROF
KINGJEROBOAMBEGANABIJAHTOREIGNOVERJUDAHTHREEYE
ARSREIGNEDHEINJERUSALEMANDHISMOTHERSNAMEWASMIC
AIAHTHEDAUGHTEROFURIELOFGIBEAHANDTHEREWASWARBE
TWEENABIJAHANDJEROBOAMANDABIJAHJOINEDBATTLEWIT
HANARMYOFVALIANTMENOFWAREVENFOURHUNDREDTHOUSAN
DCHOSENMENANDJEROBOAMSETTHEBATTLEINARRAYAGAINS
THIMWITHEIGHTHUNDREDTHOUSANDCHOSENMENWHOWEREMI
GHTYMENOFVALORANDABIJAHSTOODUPUPONMOUNTZEMARAI
```

Verses in Matrix

The Second Book of the Chronicles 12:11-13:4

A keen interest in all things esoteric is also indicated with the number 25/7 and corresponding card. Diana was reportedly receiving tuition on how to expand and develop her spiritual awareness and healing potential. Her desire to do this can clearly be seen in the photos of her with very ill or badly injured and disfigured children. (Please see photo elsewhere in this publication).

The name of "Costner" must be broken into syllables to be found in the matrix above; for some, lessening its power. Likewise, the film title "The Bodyguard". "Diana and I had been talking about doing Bodyguard 2. I told her I would take care of her just the same way that I took care of Whitney. She wanted me to write it for her. I said, 'I'll tailor it for you if you're interested'. She goes, 'I am interested'." Costner is said to have received the first draft of the script for The Bodyguard 2 on August 31st 1997, the day Princess Diana died. Reference:http://www.digitalspy.com.au/movies/news/a375832/princess-diana-wanted-bodyguard-2-role-says-kevin-costner.html

Sadly enough, it now seems, if the Bible Code is true, Diana may have had the starring role in her own movie, a remarkable remake of British Intelligence.

There are some things (which, over the course of a lifetime, **may function negatively**, on Diana's numerology chart), I will specifically point out. Diana's name at birth:

D I A N A   F R A N C E S   S P E N C E R
4 9 1 5 1   6 9 1 5 3 5 1   1 7 5 5 3 5 9
= 20   = 30   = 35

= 85 = 13/4 DESTINY, aggregate of the vowels and consonants, which equates with the **Death Card,** discussed elsewhere.
Diana's and Dodi''s Mercedes hit the **13th pillar** in the Alma Tunnel.

Inner Self, Soul or Value of the Vowels = **27/9**, rules birth and a fresh new wave, equates with the **Ace of Wands**.

Personality Projection to the World or Value of the Consonants = **58/4** The **Four of Swords,** showing a Knight lying on his tomb, the loser in a duel. Sometimes, this can mean the death of a suitor. It also can mean banishment or exile from one's homeland, due to violent altercation.

DIANA SPENCER
20 35
= Destiny Master 55/1 equates with the **Ace of Swords**, the highest elevation of the Number 1, the day on which Diana was born. This number rules the birth of a very special child, who will in some way bring light into the world. It brings engagement and marriage of honour, as the crown

indicates. It also rules the birth of a child of mixed race. It is quite possible that Diana saw this move as a way of healing racial and religious discrimination in the world.

Inner Self, Soul or Value of Vowels = **21/3 The World** card, discussed elsewhere.

Personality Projection to the World or Value of Consonants = **34/7 Eight of Wands,** which I call "the stings and arrows of poisonous society" discussed elsewhere in this publication. This number can bring serious emotional problems as a result of estrangement coupled with lack of understanding from family and friends. As a result, depression may result. Often the choice of partner is derided and even abhorred. In exaggerated cases, murder is a possibility.

LADY DIANA
3147 20
= 15 20
= 35/8
19/1
16/7

35/8 Equates with the **Nine of Wands**. This number is known as the Inheritance 8, and inherits money, power, influence, position in society, often suddenly, all at once. They are tailor-made for the job, and often step into the position someone else has vacated, taking over their place. Diana's sister Sarah was Charles' previous girlfriend, and many at the time thought he would marry her. The same would also be true of Camilla.

Inner Self, Soul Value of Vowels = **19/1 The Sun**, likes to be the boss, at the head of things. Often produces male children exclusively. The face is the most pronounced body part of this person. "The sun maketh the face to shine."

Personality Projection, Value of Consonants: **16/7 The Tower** as discussed elsewhere in this book.

Diana, Princess of Wales, was commonly known as Princess Diana.

PRINCESS DIANA
79953511 49151
= 40 20
= 60/6 Destiny
25/7 Soul
35/8 Personality

60/6 Destiny equates with the **Six of Swords** and often rules emigration to another country, and marriage to a man of foreign birth. It also rules second marriage with children involved. The man steers the woman and child to a new home on foreign shores. This number also rules a love of boating and yachting, fishing etc., and the woman will be attracted to a man who lives his life that way. This picture fits Dodi Fayed and the Jonikal romance entirely.

Soul**: 25/7** – this card and number dealt with elsewhere. **Knight of Wands.** Diana's Life Lesson, Spirit Force number vibration.

Personality Projection: **35/8, Nine of Wands**, dealt with above. Destiny of "Lady Diana".

Prince William and the connection with the loss of his mother Princess Diana in young adulthood, through his birth chart.

Prince William has the number 50/5 ruling his life from birth until the age of 27. (Add the value of the first three letters of his name to the month he was born: WIL+6 = 23+9+12+6=50/5). The card symbolizing this number is often called The Wish Card, and offers fulfillment of your hopes, dreams and desires. A couple in happy embrace can be seen, with two children dancing happily at their side. However, the negative of this card represents a very different picture. It brings dissension and arguments within the family home, the direct opposite of what the positivity of the card implies. Look to the sky, where a beautiful rainbow shines above. This represents tears of sorrow falling for the loss of a loved one. The teardrops ascend to heaven to form the multi-hued rainbow in the skies. If you look at the chart of Princess Diana, you will see she had this number and this card, ruling her life from age 54 to 81. She never got to live through those years. This card also brings a child named in memory and honour of the loved one who has passed.

As we would expect with an heir to the throne, Princes Charles and William both have the same Life Lesson or Spirit Force: Master 47/11 as symbolized by the Seven of Cups. What they came here to learn and do in this lifetime is identical. The number 43/7, represented by the Three of Cups, indicates that he will probably be surrounded by 3 women, more than likely two daughters. One of them will have the name "Mary" as her first or second name, as this number I believe represents the three Mary's in The Bible: Mary Magdalene, Mary Martha, and the Virgin Mary, mother of Jesus. Catherine Elizabeth is the name of William's wife, commonly known as "Kate", the Duchess of Cambridge. Frequently, the name "Mary" crops up either in the name of the wife or the daughters of a man with this number on his chart. I believe that both the names of "Diana" and "Mary" will be part of, if not the full name chosen for William's and Kate's first daughter. As Kate's second name is

Elizabeth, the name of William's paternal grandmother, Queen Elizabeth II, this name too will probably be included.

The number 52/7 ruling William's life between the ages of 54 and 81 indicates a man who lives his life as much out of the glare of the spotlight as he can. He may separate himself from his usual way of life at some point. He prefers the mountains or the seaside where the air is uncontaminated and pure. He is reserved, refined, and prefers the company of similar gentlemen and ladies. He will in all probability become a widower during these years, and never remarry.

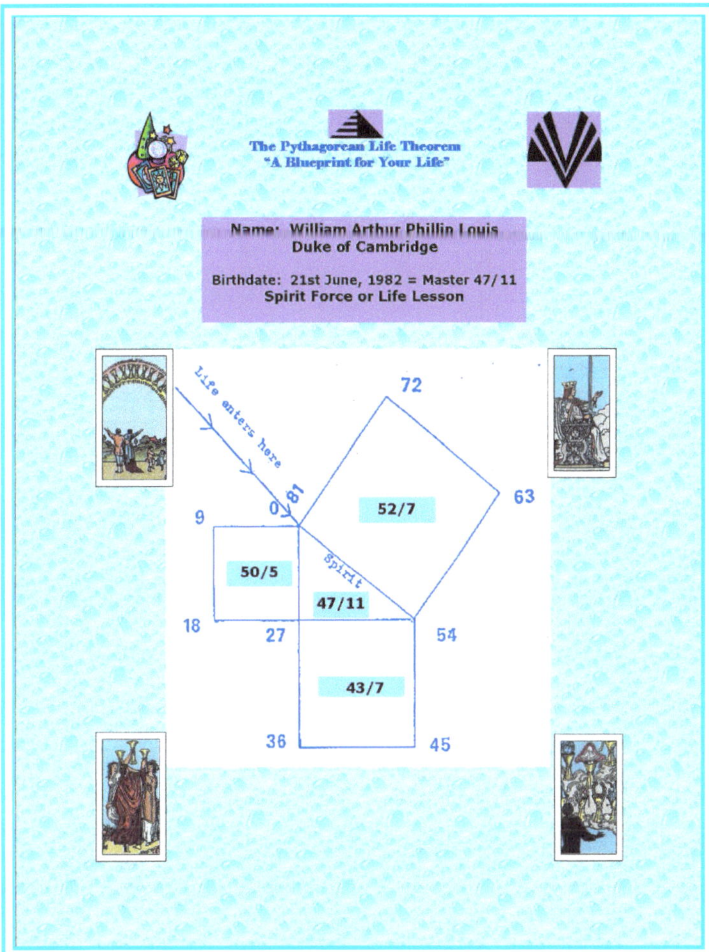

Section 32. Bad Boy Fayed

The 2012 Olympics are about to be held in London, from 27th July to 12th August, and to celebrate, they're doing a remastering of Chariots of Fire. With the second French inquest into Diana's death currently held in abeyance, in a right honourable rush to distance himself from the Fayeds, film producer Lord David Puttnam, shot the film's new release in the foot with this announcement to the press. "I threw Dodi Fayed off the Chariots of Fire set for trying to give the cast cocaine." Dodi Fayed was Executive Producer, and his father

Mohamed Al Fayed the bankroller of Chariots of Fire, Academy Award Winner for Best Picture, (in total 4 Academy Awards), at the 54th Academy Awards in 1982. Held five years after Diana's and Dodi's deaths, it had been 13 years since the Brits won the Academy's top honour. The win was and is still today, a somewhat controversial one.

Reference: http://en.wikipedia.org/wiki/54th_Academy_Awards

Puttnam's role as Producer of Chariots of Fire was not without clashes with the Fayeds. Yet he at the time publicly acknowledged that the film would not have been made without the Fayed backing. In the same breath that he now posthumously trashes Dodi's reputation as "lazy and drug-addled, a disappointment to his father, with the attention span of a flea", he admits in this interview below that Mohamed Al Fayed mistakenly gained "the impression that there was some sort of malpractice afoot" during the making of the film. This misunderstanding he attributes to the overzealousness of Mohamed's lawyer, who was, he seems to insinuate, partly responsible for influencing Mohamed in financially backing the film.

Films run over budget all the time. That's one of the reasons they have legally-binding completion guarantees. Puttnam stated that no one asks if a successful film was over budget, so no one should complain if they run over budget for the London Olympics. Although the film was made in 1995, two years before Dodi died, it did not become a world-wide hit until 1981, some four years after his death; yet Puttnam gives no credit at all for that success to Dodi's posthumous fame gained in dying at Princess Diana's side, conjoined with his role as Executive Producer for the film. The curiosity factor alone would have had enormous box office pull. Ally that to the controversy that arose in the wake of the deaths of the couple, rumoured to have been expecting a child and about to become engaged, and you just might have a global hit on your hands.

Whether or not the film became a box office flop or proved a resounding success, it doubtless had intrinsic worth. Yet, despite the widely acknowledged philanthropic credit given to Mohamed Al Fayed by other critics, neither he nor Dodi are given any credit for having an inkling, let alone an understanding or appreciation of, the worth of this film. The original sound track of the film alone has become an instantly recognizable classic piece of music. To hear Puttnam talk, both Fayed men acted like dimwits.

Puttnam paints a picture of Mohamed doing his lawyer a personal favour in bankrolling the film, while simultaneously valiantly but foolishly providing his good-for-nothing son with a career in the film industry. The nameless lawyer was evidently an old school friend of Liddell. All was well until, Mohamed (allegedly, and without foundation) suspected his lawyer of fraud. If the said lawyer was instrumental in persuading Mohamed to fund the now famous project, and he was entirely innocent of any wrongdoing, then it was hugely remiss, not to mention disrespectful, for neither the journalist Barber nor Puttnam to name him in the article.

Mohamed is said to have been convinced the film was worthless because "he felt he was being railroaded into investing in a doomed project about two sprinters with no sex, no beautiful women and not a fast car in sight." Why on earth would he fund it at all, if that was the foundation and sum of his motivation? Surely the man didn't amass a fortune by being a gullible, vacuous idiot, given to indulging his layabout drug- addicted son, as Puttnam would now have us believe. He'd surely be more savvy?

If Dodi did provide cocaine to those on the film set, as Puttnam alleges, to speak about him posthumously in such a degrading manner will not do the film's rating at the Olympics any good at all. After all, I assume both Dodi's and Mohamed Fayed's names are still on the credit roll? Furthermore, I could not imagine Princes William and Harry being exactly pleased with Puttnam making such disparaging remarks about their mother's former partner. After all, it is no secret that Dodi and Diana were having a serious sexual relationship. Paul Burrell has also reported that Diana knew Dodi used cocaine. If it was not their intention to cheapen her memory, by lowering his reputation to gutter level, inadvertently and probably unintentionally, they unfortunately managed to do a very good job of it.

David Puttnam was appointed Commander of the British Empire (CBE) in 1983, knighted in 1995, (two years prior to Dodi's death) and created a life peer on July 12th 1997, (one month prior to Dodi's death and two years after making the film) as Baron Puttnam of Queensgate, in the Royal Borough of Kensington and Chelsea.

Here's the reportage..... Quote: RICHARD BARBER
PUBLISHED: 21:14 GMT, 14 July 2012 | UPDATED: 21:18 GMT, 14 July 2012

Film producer Lord Puttnam has revealed he ordered Dodi Fayed off the set of Chariots Of Fire after the man who later stole Princess Diana's heart tried to give the cast cocaine. Dodi was an executive producer on the Oscar-winning film – due for re-release this week – after his billionaire father Mohamed Al Fayed set him up in the movie business. But according to Puttnam, Dodi was more interested in drugs and girls than in the film – which told the story of the 1924 Olympics.

Executive producer Dodi Fayed, left, was banned from the set for providing the crew with cocaine by producer David Puttnam, right. Extreme right:

Puttnam as he is today: The Right Honourable The Lord Puttnam Kt, CBE, FRSA.

'Dodi had other things on his mind than developing a film career for himself, of which girls and drugs rated pretty highly – and not necessarily in that order,' he said. Puttnam arrived one day to find a number of cast members whose mood had clearly been altered by what he suspects was cocaine provided by Dodi. 'I said to Dodi, "With the best will in the world, Dodi, this didn't happen. And I never want to see you again around my cast and crew." It was very unpleasant,' said Puttnam. The story was leaked to Dodi's father. 'It upset him terribly,' said Puttnam. 'Al Fayed had spent a huge amount of his life dealing with his son's problems and trying to contain them. 'So maybe he found it difficult now that it was out in the open. It was all very sad.'

Though he will always be remembered as the man who died beside Princess Diana in a car crash in Paris in 1997, Dodi was a little-known playboy at the time of the film's making in 1981. Dodi Fayed 'disappointed' his father Mohamed Al-Fayed, when he found out about his son being banned as had created the company to give his son new priorities in life other than drugs and women. Al Fayed had been so desperate to tame his errant son that he had set up the film production company Allied Stars in a vain attempt to carve out a career for him.

It was this company that provided nearly £20 million to finance the early development of Chariots Of Fire, meaning its future hinged on the whims of an unreliable and often drug-addled young man. Puttnam said: 'He was one of the laziest human beings I've ever come across. So the idea of him being an executive producer was always going to be hopeless because he had the attention span of the average flea.'

On the brink of the London Olympics, Puttnam has revealed other dramas behind the shooting of the movie. It told the story of Eric Liddell, the Scottish sprinter and devout Christian who withdrew from the men's 100m sprint after learning the heats fell on a Sunday. Though the 100m was Liddell's best event, he trained instead for the 400m – which he famously won.

Puttnam refers to the film as his 'proudest achievement'. His unlikely partnership with the Fayeds was forged by chance. Puttnam hit on the idea of working on a film about the Olympic Games, inspired by his father Len who had worked as picture editor at the Associated Press news agency and was responsible for images during the 1948 Olympics. During research Puttnam stumbled upon Liddell's story. It just happened that Al Fayed's lawyer at Allied Stars had attended Eltham College with Liddell and became determined to make a film about his hero.

Chariots of success: The film became a world-wide hit in 1981 and has now been digitally re-mastered for the cinemas. A scene from Chariots of Fire shows Nigel Havers, Daniel Gerroll, Nick Farrell and Ben Cross carrying Ian Charleson on their shoulders.

'So enthusiastic was the lawyer about the project that it all got nasty at one stage because Al Fayed began to suspect, and quite wrongly, that there was some sort of malpractice afoot,' said Puttnam. 'He felt he was being railroaded into investing in a doomed project about two sprinters with no sex, no beautiful women and not a fast car in sight.'

The 16-week shooting schedule was pitted with even more disasters. For Puttnam, 'the beating heart of the film' was lead actor Ian Charleson who played Liddell. 'Ian had read the Bible inside out. He absolutely embodied what Liddell was about and he had the undying respect of every single person on that movie. He was the core of the project.' So when Charleson came close to being severely injured, Puttnam also came dangerously close to scrapping the film altogether. He said: 'Liddell is prevailed upon at one stage to run in a field. We found one which was very beautiful but no one had checked the ground, which turned out to be riddled with rabbit holes. If Ian's ankle had snapped in two? Well, end of film.'

Meanwhile co-star Nigel Havers broke a wrist while practising the hurdles. But the actor was so fearful he would lose his part, he bound his arm and kept the injury a secret. And Ben Cross, cast as Jewish athlete Harold Abrahams, quickly became 'difficult to work with', according to the director.
Limited funds also meant there was no budget to pay the 7,000 extras required for the Olympic scenes. To attract crowds, Puttnam came up with the idea of an hourly prize draw, where vacuum cleaners, washing machines, motorbikes and a Ford Fiesta would be given away at the set in Birkenhead.

When the movie opened in 1981, there were mixed reviews and poor audiences at the single London cinema where it was first shown. 'I was on holiday in Cornwall and phoned the box office every two hours,' said Puttnam. 'The second afternoon performance, I was told, had been only one-third full. "Well, that's it," I thought. "It's over." 'But then an extraordinary thing happened. Audience figures peaked the longer the film was showing. In America, audiences got younger and younger as the attendances grew. Slowly, we had a global hit on our hands.' Eventually Chariots Of Fire took

£26 million at US box offices alone – and Al Fayed made a reported £6.5 million. End of Quote.

Reference: http://www.dailymail.co.uk/tvshowbiz/article-2173588/As-film-released-producer-David-Puttnam-reveals--I-threw-Dodi-Fayed-Chariots-Fire-set-trying-cast-cocaine.html#ixzz211j0S1Af

Let us see what Mohamed Al Fayed has to say about how he came to be involved in the making of Chariots of Fire. "A script had been knocking around for years at a production company called Goldcrest. The title on the cover was Chariots of Fire. Dodi took the script to his father. He persuaded Mohamed Al Fayed to put up the money to make the film. He did not need to do much persuading. As soon as Mohamed Al Fayed read the script, the story of one man who will not compromise his conscience but still wins an Olympic Gold Medal and another man who overcomes anti-Semitism to triumph in the 100 metres, he knew it would be a huge success."
Reference: http://www.alfayed.com/dodi-and-diana/dodi.aspx

Matches for: Putt nam Fayed made film over budget win award

Match: 1 of 1 Skip to Match

ELS (Equidistant Letter Spacing) Matrix

```
NSBETOOSTRONGFORMETHENTHOUSHALTHELPMEBUTIFTHEC
HILDRENOFAMMONBETOOSTRONGFORTHEETHENIWILLCOMEA
NDHELPTHEEBEOFGOODCOURAGEANDLETUSPLAYTHEMANFOR
OURPEOPLEANDFORTHECITIESOFOURGODANDJEHOVAHDOTH
ATWHICHSEEMETHHIMGOODSOJOABANDTHEPEOPLETHATWER
EWITHHIMDREWNIGHUNTOTHEBATTLEAGAINSTTHESYRIANS
ANDTHEYFLEDBEFOREHIMANDWHENTHECHILDRENOFAMMONS
AWTHATTHESYRIANSWEREFLEDTHEYLIKEWISEFLEDBEFORE
ABISHAIANDENTEREDINTOTHECITYTHENJOABRETURNEDFR
OMTHECHILDRENOFAMMONANDCAMETOJERUSALEMANDWHENT
HESYRIANSSAWTHATTHEYWEREPUTTOTHEWORSEBEFOREISR
AELTHEYGATHEREDTHEMSELVESTOGETHERANDHADAREZERS
ENTANDBROUGHTOUTTHESYRIANSTHATWEREBEYONDTHERIV
ERANDTHEYCAMETOHELAMWITHSHOBACHTHECAPTAINOFTHE
HOSTOFHADAREZERATTHEIRHEADANDITWASTOLDDAVIDAND
HEGATHEREDALLISRAELTOGETHERANDPASSEDOVERTHEJOR
DANANDCAMETOHELAMANDTHESYRIANSSETTHEMSELVESINA
RRAYAGAINSTDAVIDANDFOUGHTWITHHIMANDTHESYRIANSF
LEDBEFOREISRAELANDDAVIDSLEWOFTHESYRIANSTHEMENO
FSEVENHUNDREDCHARIOTSANDFORTYTHOUSANDHORSEMENA
NDSMOTESHOBACHTHECAPTAINOFTHEIRHOSTSOTHATHEDIE
```

Verses in Matrix

The Second Book of Samuel 10:11-10:18

Section 33. Dr Hasnat Khan, heart surgeon, cousin of Imran Khan, and his relationship with Princess Diana.

Described as 'Mr. Wonderful' by the late Princess Diana, The Princess of Wales, Dr Hasnat Khan was said to be the 'love' of the deceased Diana's life.

Khan had met Diana at the Royal Brompton Hospital, where she was visiting a friend recovering from heart surgery. They pursued a discreet two-year affair that finished a few months before her death in a car crash in 1997. Diana's friends are reported to have described Khan as the 'love of her life' and to have spoken of her distress when he ended their relationship. She dubbed him 'Mr. Wonderful.' It has been claimed that Khan decided not to marry the Princess of Wales because of their cultural differences.

Khan, a Muslim, ended his relationship with the Princess only months before her death after concluding that a marriage between them would be doomed to failure. Dr. Khan told his family: "If I married her, our marriage would not last for more than a year. We are culturally so different from each other. She is from Venus and I am from Mars. If it ever happened, it would be like a marriage from two different planets."

The inquest into the Princess's death heard evidence last week from one of her closest friends, Rosa Monckton, that Diana had no plans to marry Dodi Fayed, who died with her in a Paris car crash 10 years ago, and that she was still infatuated with Dr. Khan. Ms. Monckton said the Princess had been "deeply upset and hurt" when Dr Khan broke off their relationship in the summer of 1997. "She was very much in love with him. She hoped that they would be able to have a future together. She wanted to marry him," she told the hearing.

Rosa Monckton, friend of Diana. Diana in Pakistan.

Dr. Khan had a two-year relationship with Diana from 1995. She stayed at his Chelsea flat in London, and he stayed at her home in Kensington Palace. However, they always tried to keep their affair discreet and he has never talked publicly about his deep feelings for her. It is understood that at one point the Princess was willing to convert to Islam in order to marry him but abandoned the idea when he took the decision that their relationship could not work in the long term.

Other lovers, notably former Household Cavalry officer James Hewitt, may have cashed in on their relationship, but the intimate moments Dr. Khan enjoyed with Diana will never be shared with a wider audience. When in Pakistan, he lives in his spacious yet modestly decorated colonial-style house in Jehlum, a two-hour drive from the Pakistani capital of Islamabad (the site of the capture of Osama bin Laden incidentally).

Dr Khan was interviewed as part of the Lord Stevens' inquiry, which concluded that the car crash was an accident. The surgeon told officers that the Princess had considered moving to Pakistan after he had told her it was the "only way he could see them having any sort of a normal life together". The report added: "He knew that the Princess considered this an option for a while. She even went to Pakistan to speak to Jemima Khan about life there. However, after a while she no longer talked about Pakistan and although she did not actually say so, Hasnat Khan got the impression that she no longer considered living there a possibility."

Dr Khan also revealed to the inquiry that, at the end of 1996, Diana asked Paul Burrell, the former royal servant, to talk to a priest about them marrying in secret. The report said: "Mr. Khan thought it was a ridiculous idea. He believed that she wanted to get married but had not thought about the implications. They had an argument about it." End of Quote.

Reference: http://www.telegraph.co.uk/news/newstopics/diana/8874787/Dr-Hasnat-Khan-the-heart-surgeon-who-Princess-Diana-dubbed-Mr-Wonderful.html

Cultural differences were reportedly not the only reason Dr. Khan ended his two-year relationship with Princess Diana. He expressed his concern over the fact that he had spent a lifetime building a serious respect for himself as a top heart surgeon, and was not about to see his life changed radically with the media attention that would surround him constantly, which he described as "hell", should he marry Diana.

In short, he was worried about the media circus that would follow the pair constantly, and was not prepared to change his life to that extent, to marry her. He simply did not love her enough to make that kind of sacrifice; she was mistress, not wife material. Diana realized this, and it broke her heart. It was a cruel rejection. ((Please see my comments elsewhere in this publication, on the Master 33/6 year of emotional crucifixion, the Christ vibration, (since Christ was crucified at the age of 33), that precedes the 34/7 year Diana was in when she died.))

Yet it is eminently possible to fall in love in the wake of a disastrous love affair and such a relationship is often very successful indeed, long term. When Dodi entered the picture, this time it was different. Here was someone who lived, worked and moved in the world of media: a film producer. He had no such concerns as had Dr. Hasnat Khan. Here was someone who would willingly and eagerly amalgamate her life with that of his. A whirlwind courtship ensued, and if the Bible Code is true, one that culminated in Diana becoming pregnant.

It is clear in the touching photographs of the pair in their final hours that Diana had fallen in love with Dodi Fayed, and he with her. They were obviously devoted to each other. Their body language speaks loud and clear, and tells us that she leaned on him for support, and he was more than capable of providing it for her. Her right hand reaches around behind her, to cover his left hand with which he guides and keeps her close to him. If that is not a picture of a woman who loves a man, and what's more, a man who not only loves a woman, but desires to shield and protect her, then what is?

Any concerns the establishment might have had about Diana converting to the Muslim faith that possibly augured her harm, were already on the boil during her relationship with Dr. Khan; long before Dodi entered the picture. Some newspaper reports suggested that Doctor Hasnat Khan was the father of Diana's child, even though he testified under oath that Diana took the contraceptive pill while with him. The pill is not foolproof, so I decided to search for the possibility of Diana carrying Hasnat Khan's child.

A distressed Paul Burrell, Diana's butler, outside the inquest. Paul was a very significant part of Diana's life. She relied on him for support and help, both physical and emotional, on a daily basis. His wife Maria Cosgrove, was Diana's dresser. Yet it would seem, on those matters of political importance in which she was involved, like the land mines campaign, she turned to Simone Simmons, her spiritual adviser, in preference to Paul or any member of her family to guard her files. She did however, entrust Paul with secret documents outlining her fears for her safety. Perhaps she, very wisely, decided not to put all her eggs in the one basket.

Will the real Diana please stand up? This picture is to my mind so typically the real Diana. The Master 22/4 agent is boyish, and looks her best in male influenced suits, three quarter page boy style pants and hairdo's. Being the mother of two boys, she was in her element. If The Bible Code is true, within months she was to deliver another fine healthy boy.

Quote: The mother of Princess Diana called her a 'whore' for dating Muslim men, her inquest has heard. Frances Shand-Kydd made the 'disgraceful' comment when she discovered her daughter was in a serious relationship with heart surgeon Hasnat Khan. The pair did not speak again before Diana died two months later, according to her butler Paul Burrell.

Reference: http://www.dailymail.co.uk/news/article-508191/Dianas-mother-called-whore-sleeping-Muslim-men.html#ixzz20TC5qAUs (15th January 2008)

The sensational revelation came on the day that he told how:

- Diana was planning to marry Dr Khan;
- Her romance with Dodi Fayed was just a '30-day' fling;
- Prince Philip did write 'cutting' letters to the princess but would not have ordered her murder;
- The Queen warned him of mysterious 'powers at work' - but he had no idea what she meant.

The 49-year-old former butler revealed that Diana's bitter conversation with her mother happened in June 1997, during the last throes of her relationship with Dr Khan and just two months before she and Dodi died in Paris.

Mr Burrell told the High Court that the princess had held up the phone as they sat together on the sofa of her Kensington Palace apartment so that he could hear her mother's rant.

He said Mrs Shand-Kydd, who died in 2004, was a 'formidable lady' who often expressed herself 'in extremely forceful terms about Diana's consorts, 'especially if they were Muslim'.

Asked to describe what he had heard on that particular day, he hesitated. It was only when the coroner, Lord Justice Scott Baker, interjected, saying: 'This is relevant', that Mr Burrell replied: 'Well, she called the princess a whore and she said that she was messing around with f****** Muslim men and she was disgraceful. She said some very nasty things.'

Mohamed Al Fayed's barrister, Michael Mansfield, asked: 'It was shortly after one of these telephone calls that the princess decided she did not want to talk to her mother again?' 'Yes,' Mr Burrell confirmed.

Mrs Shand-Kydd went through a bitter divorce from Diana's father Earl Spencer, after she left him for wallpaper tycoon Peter Shand-Kydd. She endured a rocky relationship with her youngest daughter and had no idea how serious Diana's relationship with Dr Khan was. In fact the princess had been planning to marry the man she described as her 'soul mate' just months before she embarked on her high-profile relationship with Dodi Fayed.

Mr Burrell told the court that he discussed the matter in confidence with Father Anthony Parsons, a priest from a Catholic church near Kensington Palace, without Dr Khan's knowledge. But the pair split up the month before Diana died in a clandestine night-time meeting in Battersea Park, South London.

As he took the stand yesterday, Mr Burrell appeared nervous. His voice was barely audible and he failed even to remember the date of his wedding with wife Maria, who worked as Diana's dresser. As he went on, however, he gave a voluble account of his time with Diana and her troubled personal relationships, describing Dr Khan as the love of Diana's life.

'The princess said that this was her soul mate, this was the man she loved more than any other and it was a very deep spiritual relationship,' he said. 'I witnessed it at first hand and they were very much in love.' While she quickly became fond of Dodi, the relationship was dismissed by Mr Burrell yesterday as a '30-day' fling. Diana's highly public romance, he insisted, was part of her attempt to 'get back' at Dr Khan.

Asked why she would want to humiliate a man she had cared for so much, Mr Burrell replied: 'I don't think that a relationship that lasted for 18 months was gone overnight. The princess was still burning a candle for Mr Khan. She was still in love with him.'

"I could have saved Diana," wept Hasnat.
Last updated at 22:00 02 September 2006
http://www.dailymail.co.uk/news/article-403364/I-saved-Diana-wept-Hasnat.html

Quote: "I can't find him. I can't find him!" There was a note of pleading in her voice. The Princess was sobbing uncontrollably in the sitting room, clutching a handful of soggy tissues. "I don't know where he is," she said, as she perched on the edge of the sofa, rocking backwards and forwards in her white towelling robe, as vulnerable and as upset as I had ever seen. My heart went out to her.

It was the night of September 15, 1996, 18 days after the divorce had become absolute. It was nearly midnight and I had been summoned to Kensington Palace by a frantic, tearful call. She was alone. The telephone, with its long, snaking cord which allowed her to walk from room to room, was at her side. She didn't need to say much. We both knew what this was about: Dr Hasnat Khan, the man she was truly in love with, the man whom her inner circle called The One. It was a year since they had met in 1995.

"I've tried to contact him for five days now, and left messages everywhere. What more can I do?" she said. "I've been to his flat - it's in darkness. I've left messages on his bleeper and at the switchboard, and I don't know where he is."

Hasnat was an eminent heart surgeon at the Royal Brompton Hospital in Chelsea. Whenever she left a message, she used the codename Dr Allegra. She also had a secret nickname for him: DDG - Drop Dead Gorgeous. But that night he wasn't answering to any name. I tried to offer reassurance, although I knew there had been a minor altercation in a telephone conversation the previous week.

"Can I ask you to go out and find him and deliver him a message?" she asked. She knew I'd say yes. She could hardly have gone out trawling the streets herself. After visiting his flat, his local pub and a takeaway he used, I eventually found him in a wine bar. Sat in the corner at a table filled with empty beer bottles and amid smoke that clung to your clothes, Hasnat smiled, as if my visit was reassurance for him, too, that he was being missed.

"Join me for a beer, Paul," he said. Over a drink, I delivered the message I had been given and he gave me one to take back to the boss. It was after 1am when another cab dropped me off at Kensington Palace, or KP as we called it. As I walked up the cobbled passageway and into the back of Apartments 8 and 9, the boss must have heard my footsteps. "Did you find him?" she asked before scurrying me into the sitting room. I delivered the message. The face that had been so wretched a couple of hours earlier lit up with joy. She was in love again.

Reporting for work a few hours later, I found a note in the pantry. It read: "Dear Paul, Not many people would venture out late at night to sort a heart out...I am profoundly touched by your actions last night...Love from Diana." With Hasnat, she was the happiest I had ever seen her.

In the weeks before her death, I had read about Diana and Dodi with dismay. I remember the photo of her wearing white shorts over a black body and carrying her deck shoes, with Dodi walking behind her, and thinking she was not happy. I also recall her ringing me from the yacht. "I'm ready to come home now," she said, fed up with the summer fling.

Had the outside world known the happiness she exuded when she was with Hasnat, it would have understood that what she experienced with Dodi was not love, nor was it serious. I am convinced that those photos with Dodi were signals sent out to grab the attention of the man she really loved.

It was September 1995 when the Princess first laid eyes on Hasnat, in one of those classic moments when lift doors are about to close and someone puts a foot between them. She was at the Royal Brompton Hospital and the foot belonged to Hasnat. "My heart nearly stopped!" she laughed.

It wasn't long before she fell deeply, and unexpectedly, in love. She adored his hands, his large brown eyes, his long eyelashes. "Can you imagine what beautiful children we would make?" she winked.
He always felt he wasn't good enough. What he failed to understand was that he was all she ever wanted - a man offering nothing but himself. There was no underlying motive. Indeed, he preferred to be a secret.

He was a heavy smoker who liked a good drink. The boss was immaculate, loathed smoking and obsessed with keeping trim. He felt she could have had anyone in the world, but in all their time together, she never looked at another man. I knew it was serious when she started buying ashtrays. Each morning, I'd find ash in the dustbin in my pantry, on top of Kentucky Fried Chicken boxes.

Hasnat knew the routine to get into KP. I would pick him up in the BMW in a dark corner of a cobbled, quiet mews, an agreed meeting place. As we approached the palace, he'd climb into the back seat, grab the tartan rug, lie flat and pull it over himself. The policemen on the gates waved me through.

Instead of driving to the front door of Apartments 8 and 9, I turned sharp right into King's Court, the route to Princess Margaret's front door. This was the ideal spot to usher inside secret guests because the Queen's sister had insisted no CCTV cameras were to monitor her front door. The boss wasn't the only Princess intent on privacy.

The sound of a car engine always brought a twitch from Princess Margaret's curtains at Apartment 1a opposite. "She just doesn't miss a trick!" the boss would laugh. The back door had been left ajar and the Princess was standing there. After letting Hasnat in, she made an exaggerated wave to the windows opposite. "Night, Margo!" she would shout.

As the relationship intensified in the summer of 1996, the boss said: "I want him to spend more time here - to come and go as he pleases." She suggested using a spare room on the ground floor as a den for Hasnat. "We could find new sofas, put in a widescreen TV so he can watch football - even install a bar and a fridge. William and Harry could join him at weekends," she imagined. The fact she saw the man in her life integrating with her sons explains how much in love she was.

One of my duties was to ensure the relationship was hidden. It changed routines like never before. Doors were closed, marked out of bounds. I told the chef the Princess was terribly hungry to explain why she wanted double portions, which I would divide and serve as dinner for two. And the kitchen must have noticed her sudden passion for bacon sandwiches with HP sauce. The usual reserves of Montrachet white wine were replaced with Budweiser in the refrigerator.

Sundays became days off for me. With the chef also sent home, she revelled in the freedom to be mistress in her home. She was never renowned for her culinary skills but started making cheese on toast, omelettes and pasta. She used the microwave, rather than the oven. And, of course, there was Kentucky Fried Chicken when she couldn't be bothered to cook.

The relationship was clandestine, treated like a shameful affair, yet the Princess was single and joyously in love. No yachts, no toys, no helicopters, and no extravagant purchases. This was the real thing.

She would disappear to spend whole days in Hasnat's one-bedroom flat. Diana, Princess of Wales, would return to the luxury of KP, sit down with a glass of carrot juice and talk, as if she was an ordinary woman, about how she had spent the day in a pokey flat, vacuuming, dusting, doing the dishes, ironing, making his bed. She craved such normality away from the trappings of Royalty which had both defined and stifled her.

It was around December 1996 when she said it. And, unlike most things, I hadn't seen it coming. "This is the man I want to marry," she declared, curled up on the sofa in her sitting room. "Do you think it would be possible for me to marry in secret with just a couple of witnesses?" I was waiting for her to fall about laughing. But she was serious and was waiting for an answer.

"I'm not sure what the official protocol would be," I said. "But I do know someone I could sound out discreetly." The Burrell family priest, Father Tony Parsons, sprung immediately to mind. He was completely trustworthy. We had an extraordinary conversation sitting in the pews of an empty church. He looked aghast when I asked. "It's impossible, Paul! Impossible! It could never be done." So, I was the bearer of bad news on my return to KP. The Princess was disappointed but understanding and never spoke of it again.

I spent many evenings with Hasnat, mainly at the Anglesea Arms in Chelsea, talking about the Princess over pints of Guinness. The next morning, at breakfast, she would ask: "Tell me! What did he say?"

By May 1997, the Princess didn't want to hide him any longer. And she knew how to go about it: she would set up a photo opportunity of them together. But when her master plan was suggested to Hasnat, his reaction was not what she had expected. "I'm just not ready for it," he told me.

He dug in his heels and it upset the boss. "I've had it with him!" she said. The row led to a difficult and strained June, and matters came to a head in the first week of July. She arranged to meet Hasnat on neutral territory, at a dimly-lit spot. They drove there separately. The relationship was to end as clandestinely as it had been conducted. It has long been reported that Hasnat ended it. The truth is that the Princess did. She ended it harshly, and in tears, as they stood in the middle of a London park.

The following morning, her hurt and anger had subsided, and she wondered whether she had done the right thing. "Do you think he'll ring me?" she asked, and I knew she had acted rashly, and was regretting her impulses. I offered to take him a message but she refused. "He'll come round," she told me. But he never did, and nor did she relent.

Within days, she was with the Al Fayeds on the Jonikal. But I am convinced Hasnat was not far from her thoughts and her public display of affection towards Dodi was her way of making him sit up, take note and realise what he could lose. But the phone stayed silent, no letters were received and there were no messages for me to deliver. I think Hasnat felt he had been pushed into an intolerable situation from which there was no way back.

In the days after the tragedy in Paris, he was the only man I could think about. I didn't know whether to contact him. Then, I received a phone call. "I could have saved her," he said over and over again.

Hasnat said he knew the exact procedure to save a life threatened by a ruptured aorta. I told him what I'd learned in the Paris hospital. He pressed me for details and became more and more upset. "Why did she have to be so far away?" he said. "I could have saved her here."

It seemed futile to talk about it any further. She was dead. But we arranged to meet at the end of the week at the bottom of the KP drive. Just as I left Apartments 8 and 9, I noticed on my desk the dark blue cloth band that the Princess used to tie back her hair. Instinctively, I picked it up, slipped it into my pocket and headed for my rendezvous.

When Hasnat pulled up alongside me in his battered old car, he wound down the passenger window. He looked awful, as if he hadn't slept for days; his eyes were dark and baggy, and his hair was unkempt. "I had to see you," he said. We hardly spoke but didn't need to. As he made to leave, I said, "Wait, I have something for you." I reached into my pocket, brought out the dark blue cloth and pressed it into his hand. His eyes brimmed with tears.

He held it to his nose and mouth for a second or two, breathing in the scent of the Hermes perfume, 24 Faubourg, she used. He put it into his pocket, nodded, and tried to say something - but couldn't. "I'll see you at the funeral," I said. And he was gone. End of Quote. Extracted from The Way We Were by Paul Burrell.

Everyone seems to have their own opinion about whether or not Diana was in love with Dodi Fayed. Diana's stepmother, who had renewed her relationship with Diana after her father Charles's death, claims:

'She WAS in love with Dodi - and they would have got married' By PAUL HARRIS Last updated at 01:37 13 December 2007

Quote: Princess Diana was "madly in love" with Dodi Fayed and was probably about to marry him, her stepmother insisted yesterday. Raine, Countess Spencer told the inquest on the couple that it was "highly likely" they would have become engaged and married - or at least started living together.

But in her first detailed testimony about the relationship, she said she believed Diana was too "old-fashioned" to become pregnant outside marriage - and that she never mentioned it during their intimate conversations. End of Quote.

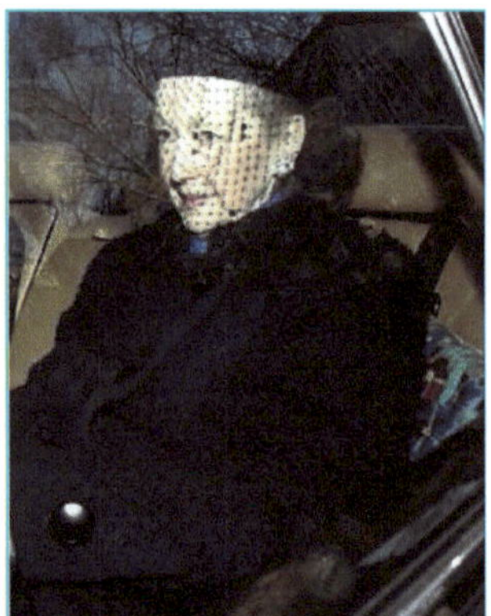

Fingers entwined, so in love. The last pictures of Dodi and Diana together give weight to Diana's stepmother Raine's testimony.

http://www.dailymail.co.uk/news/article-501520/Dianas-stepmother-She-WAS-love-Dodi--got-married.html#ixzz2181t5XXI

Was Princess Diana murdered for nearly marrying Muslim 'secret fiance' Dodi? Published August 18th, 2011 - 09:49 GMT via SyndiGate.info

A German journalist has claimed that Princess Diana was secretly engaged to Egyptian film producer Dodi Fayed before she died in August 1997. Trixi Chall, who wrote for Germany's popular Bild newspaper for two decades, told of a visit to Monte Carlo by Princess Diana and Dodi Fayed - just seven days before the fatal Paris car crash.

Chall claimed that she saw Diana pointing at and picking out a diamond-encrusted ring from an exclusive -jeweller as they strolled through the resort. Dodi later arranged for the 11,500 pounds ring - from Repossi the jeweller's Tell Me Yes range - to be sent to the shop's sister outlet in Paris. It was then delivered to his suite at the Paris Ritz hotel just hours before the couple died.

The remarkable evidence has added weight to the earlier claims made by former Harrods owner Mohamed Al Fayed that Diana and Dodi were murdered in an Establishment plot to stop the Princess marrying a Muslim. Chall has now come forward to speak for the first time about the dramatic event she witnessed during an ordinary morning shopping excursion in the swish casino resort of Monte Carlo on Saturday August 23 1997.

On that morning she went to an electrical store by the port. She said that on leaving she glanced across the road and was surprised to see Diana hand-in-hand with a man she had never seen before. "Nobody else was aware of them and they were aware of nobody, as if they were the only ones in the world. They were not in any hurry, were chatting happily and strolling casually," the Daily Express quoted her as saying.

Chall, who has detailed her account in a signed statutory declaration, added: "They went to the shop on the left of the Hermitage hotel. They looked in the window and Diana was pointing with her finger at something. I saw them entering the shop."

There has been much speculation that Diana had an engagement ring when she died after secretly -becoming engaged to Mr Fayed. Chall said she has never given her account to the various investigations because she felt strongly from very early on there was a huge "cover-up of the truth coming from the top".

Reference:http://www.albawaba.com/editorchoice/princess-diana-was-secretly-engaged-dodi-fayed-death-388913

Diana during her Dr. Hasnat Khan era. Diana was proud. Sneaking around behind closed doors would not suit her for long. Her love, based on admiration for the life-saving talents of the brilliant and dedicated Pakistani heart surgeon motivated her to try her best to fit into his lifestyle; but in truth, she needed a partner who would fit into hers. She had an important destiny of her own to fulfil, and a power play ensued. She needed a life partner who would allow her to pursue her own course in life, not hide away in his. She found that ideal man at last in Dodi Fayed. Dodi allowed her the freedom to be herself, to be open and honest about her relationship, and she fell in love anew.

Matches for: Doctor Khan not father Prin cess Diana baby

Match: 1 of 1 Skip to Match

ELS (Equidistant Letter Spacing) Matrix

```
CONSPIREDAGAINSTHIMTOSLAYHIMANDTHEYSAIDONETOAN
OTHERBEHOLDTHISDREAMERCOMETHCOMENOWTHEREFOREAN
DLETUSSLAYHIMANDCASTHIMINTOONEOFTHEPITSANDWEWI
LLSAYANDEVILBEASTHATHDEVOUREDHIMANDWESHALLSEEW
HATWILLBECOMEOFHISDREAMSANDREUBENHEARDITANDDEL
IVEREDHIMOUTOFTHEIRHANDANDSAIDLETUSNOTTAKEHISL
IFEANDREUBENSAIDUNTOTHEMSHEDNOBLOODCASTHIMINTO
THISPITTHATISINTHEWILDERNESSBUTLAYNOHANDUPONHI
MTHATHEMIGHTDELIVERHIMOUTOFTHEIRHANDTORESTOREH
IMTOHISFATHERANDITCAMETOPASSWHENJOSEPHWASCOMEU
NTOHISBRETHRENTHATTHEYSTRIPPEDJOSEPHOFHISCOATT
HECOATOFMANYCOLORSTHATWASONHIMANDTHEYTOOKHIMAN
DCASTHIMINTOTHEPITANDTHEPITWASEMPTYTHEREWASNOW
ATERINITANDTHEYSATDOWNTOEATBREADANDTHEYLIFTEDU
PTHEIREYESANDLOOKEDANDBEHOLDACARAVANOFISHMAELI
TESWASCOMINGFROMGILEADWITHTHEIRCAMELSBEARINGSP
ICERYANDBALMANDMYRRHGOINGTOCARRYITDOWNTOEGYPTA
NDJUDAHSAIDUNTOHISBRETHRENWHATPROFITISITIFWESL
AYOURBROTHERANDCONCEALHISBLOODCOMEANDLETUSSELL
HIMTOTHEISHMAELITESANDLETNOTOURHANDBEUPONHIMFO
RHEISOURBROTHEROURFLESHANDHISBRETHRENHEARKENED
UNTOHIMANDTHEREPASSEDBYMIDIANITESMERCHANTMENAN
```

Verses in Matrix

The First Book of Moses, called Genesis 37:18-37:28

Matches for: Diana die Alma Tunnel not carry Doctor Khan baby

Match: 1 of 1 Skip to Match

ELS (Equidistant Letter Spacing) Matrix

NEOFTHEPITSANDWEWILLSAYANDEVILBEASTHATHDEVOURE
DHIMANDWESHALLSEEWHATWILLBECOMEOFHISDREAMSANDR
EUBENHEARDITANDDELIVEREDHIMOUTOFTHEIRHANDANDSA
IDLETUSNOTTAKEHISLIFEANDREUBENSAIDUNTOTHEMSHED
NOBLOODCASTHIMINTOTHISPITTHATISINTHEWILDERNESS
BUTLAYNOHANDUPONHIMTHATHEMIGHTDELIVERHIMOUTOFT
HEIRHANDTORESTOREHIMTOHISFATHERANDITCAMETOPASS
WHENJOSEPHWASCOMEUNTOHISBRETHRENTHATTHEYSTRIPP
EDJOSEPHOFHISCOATTHECOATOFMANYCOLORSTHATWASONH
IMANDTHEYTOOKHIMANDCASTHIMINTOTHEPITANDTHEPITW
ASEMPTYTHEREWASNOWATERINITANDTHEYSATDOWNTOEATB
READANDTHEYLIFTEDUPTHEIREYESANDLOOKEDANDBEHOLD
ACARAVANOFISHMAELITESWASCOMINGFROMGILEADWITHTH
EIRCAMELSBEARINGSPICERYANDBALMANDMYRRHGOINGTOC
ARRYITDOWNTOEGYPTANDJUDAHSAIDUNTOHISBRETHRENWH
ATPROFITISITIFWESLAYOURBROTHERANDCONCEALHISBLO
ODCOMEANDLETUSSELLHIMTOTHEISHMAELITESANDLETNOT
OURHANDBEUPONHIMFORHEISOURBROTHEROURFLESHANDHI
SBRETHRENHEARKENEDUNTOHIMANDTHEREPASSEDBYMIDIA
NITESMERCHANTMENANDTHEYDREWANDLIFTEDUPJOSEPHOU
TOFTHEPITANDSOLDJOSEPHTOTHEISHMAELITESFORTWENT
YPIECESOFSILVERANDTHEYBROUGHTJOSEPHINTOEGYPTAN
DREUBENRETURNEDUNTOTHEPITANDBEHOLDJOSEPHWASNOT
INTHEPITANDHERENTHISCLOTHESANDHERETURNEDUNTOHI
SBRETHRENANDSAIDTHECHILDISNOTANDIWHITHERSHALLI

Verses in Matrix

The First Book of Moses, called Genesis 37:20-37:30

Matches for: Diana die Alma Tunnel baby boy Khan not father

Match: 1 of 1 Skip to Match

ELS (Equidistant Letter Spacing) Matrix

```
NEOFTHEPITSANDWEWILLSAYANDEVILBEASTHATHDEVOURE
DHIMANDWESHALLSEEWHATWILLBECOMEOFHISDREAMSANDR
EUBENHEARDITANDDELIVEREDHIMOUTOFTHEIRHANDANDSA
IDLETUSNOTTAKEHISLIFEANDREUBENSAIDUNTOTHEMSHED
NOBLOODCASTHIMINTOTHISPITTHATISINTHEWILDERNESS
BUTLAYNOHANDUPONHIMTHATHEMIGHTDELIVERHIMOUTOFT
HEIRHANDTORESTOREHIMTOHISFATHERANDITCAMETOPASS
WHENJOSEPHWASCOMEUNTOHISBRETHRENTHATTHEYSTRIPP
EDJOSEPHOFHISCOATTHECOATOFMANYCOLORSTHATWASONH
IMANDTHEYTOOKHIMANDCASTHIMINTOTHEPITANDTHEPITW
ASEMPTYTHEREWASNOWATERINITANDTHEYSATDOWNTOEATB
READANDTHEYLIFTEDUPTHEIREYESANDLOOKEDANDBEHOLD
ACARAVANOFISHMAELITESWASCOMINGFROMGILEADWITHTH
EIRCAMELSBEARINGSPICERYANDBALMANDMYRRHGOINGTOC
ARRYITDOWNTOEGYPTANDJUDAHSAIDUNTOHISBRETHRENWH
ATPROFITISITIFWESLAYOURBROTHERANDCONCEALHISBLO
ODCOMEANDLETUSSELLHIMTOTHEISHMAELITESANDLETNOT
OURHANDBEUPONHIMFORHEISOURBROTHEROURFLESHANDHI
SBRETHRENHEARKENEDUNTOHIMANDTHEREPASSEDBYMIDIA
NITESMERCHANTMENANDTHEYDREWANDLIFTEDUPJOSEPHOU
TOFTHEPITANDSOLDJOSEPHTOTHEISHMAELITESFORTWENT
YPIECESOFSILVERANDTHEYBROUGHTJOSEPHINTOEGYPTAN
DREUBENRETURNEDUNTOTHEPITANDBEHOLDJOSEPHWASNOT
INTHEPITANDHERENTHISCLOTHESANDHERETURNEDUNTOHI
SBRETHRENANDSAIDTHECHILDISNOTANDIWHITHERSHALLI
```

Verses in Matrix

The First Book of Moses, called Genesis 37:20-37:30

DETAILS of the items Diana was wearing when she died were disclosed yesterday.

The French police's inventory was produced after Paul Burrell told the inquest that when he arrived in Paris following the crash he was given an envelope containing her jewellery.

He said: 'It was quite distressing because most of the jewellery was broken, minus one earring, which was missing.'

The list of items released by the Brigade Criminelle stated that 'Lady Diana Spencer' had:

- One pair of trousers and one body - an all-in-one swimsuit-style top.
- A pair of Gianni Versace shoes (slip-on, stiletto-heeled sling-backs made of black satin) size 40.
- A black size 42 Giorgio Armani jacket, partially cut, the side pockets of which are stitched up.
- A black size 30 Ralph Lauren belt, the strap and buckle of which are made of crocodile skin.
- A gold kidney bean-shaped earring with four rolls mounted on it, slightly dented on its front edge with its clip/clasp missing, found in the Mercedes.

Six white pearls were also found in the Mercedes.

In the same verses, we get results indicating that if Diana was pregnant, Khan was definitely not the father. Paul Burrell confused the issue by claiming that Diana and Hasnat had broken up in July. If true, this did not give her much time to become six weeks pregnant to Dodi, as the other matrices indicate.

Diana, William and Harry visit a theme park, where Princes can be normal little boys having fun with Mummy.

The month Diana died. Diana died in a 34/7 year, and a 42/6 month, (34 year +8th month = 42 (4+2=6)

These are the cards for the numbers 42 and 6. They rule the union between a man and a woman. The 42 is symbolized by the Two of Cups and rules engagement and marriage in the positive. Negatively: separation, divorce and death. The couple pledges love, marriage, union by exchanging the cups of love in a ceremony. The number 6 rules their sexual union presided over by an angel. Mohamed Fayed said that they had told him of their wedding plans. The symbolism of the cards gives us no reason to doubt him.

William and Harry had spent time with Dodi, holidaying on the Jonikal yacht and skiing with him too. Worlds away from the secretive, kept-under-wraps liaison with Hasnat Khan, Diana and Dodi were able to share their life openly and publicly in company with Diana's boys. William and Harry were thus able to get to know Dodi and reportedly got on with him very well. Diana in fact wrote a letter thanking Dodi and Mohamed for the wonderful holiday she and her boys had enjoyed, and how they were having "withdrawal symptoms" back at home in Kensington Palace.

Quote: The candid, handwritten letters, on Kensington Palace notepaper, showed for the first time in Diana's words the true depths of her feelings for Dodi. In one particularly intimate letter, beginning "Darling Dodi", she thanked him for "bringing such joy into this chick's life". On a day of drama, it was also learnt that shortly before Diana died, she gave Dodi a cherished pair of cuff-links that were a gift from her father - the man she "loved most in the world". It said: "Darling Dodi, these cufflinks were the very last gift from the man I loved most in the world, my father. They are given to you as I know how much joy it would give him to know they were in such safe and special hands. Fondest love, Diana." In another letter, written on Aug 6, she spoke of her "love" for him and offered "a million heartfelt thanks" for the joy and happiness he had brought into her life. She then described a recent cruise they had taken on his father's yacht as the "most magical six days on the ocean waves". The princess said her gratefulness came "with all the love in the world". End of Quote.

Reference: http://www.telegraph.co.uk/news/uknews/1572601/Princess-Dianas-letters-of-love-to-Dodi-Fayed.html

What about Dodi's numbers at the time of his and Diana's deaths?

Dodi was born on the 15th of April, 1955, so in 1997, at his last birthday, he entered a 45/9 Year, represented by the Five of Cups. (15+4+ (1997= 1+9=9+7=26) = 45/9. Frustration in a love relationship, endings and partings often occur.

In the month of August, he was in a 53/8 month, (45+8= 53) which equates with the Knight of Swords - the number of a man in a hurry. He will move swiftly, even ride roughshod over others, in pursuit of his goal or love interest. It represents the bravest Knight in the army, who leads the charge into battle. He is unafraid, but he could be also unprepared. He might act impetuously, or be forcibly hurried in his move, with no time to mull over decisions under the circumstances. Someone catapults out of his life and just as swiftly, someone else catapults in. Since Dodi was involved with the model Kelly Fisher (and Diana with Hasnat Khan), just prior to their union, this fits the picture.

Negatively, the number 53/8 rules danger on the roads, and people suddenly coming out at you from unexpected places. I usually warn people that they might meet with someone on the highway who flouts the law or runs red lights, and to watch out for that person. They may hit you then flee. I have found in my experience this number rules hit and run accidents. Under the 53/8, you may find yourself the target of an aggressor, and your business, your bank accounts, the law and your love relationships can all be involved in this hot pursuit of you which can have legal ramifications.

Kelly Fisher claimed Dodi had proposed to her, given her a ring, then treated her very badly. She claimed that Dodi was involved with both her and Diana at the same time, unbeknown to either of them. She discovered the two-timing, she said, when she saw a picture of Dodi kissing Diana in a newspaper. In a taped phone call, Dodi denied he and Kelly were engaged, and accused her of "terrorism". In a flurry of humiliation, jealously, and revenge, Kelly was involved in suing Dodi for breach of promise when both he and Diana were killed.

Kelly Fisher wearing the alleged engagement ring Dodi gave her.

All of the above alleged events, which apply to ordinary everyday people under the number 53/8, can be seen to have applied to Dodi Fayed. By extension, they must have affected and upset Diana.

Dodi turned 42 at his last birthday. He was therefore in a 14/5 Four-Month Cycle from 15th April to 15th August, 1997. The first 4 months after your birthday are ruled by the age you are at the time.

1997
42+

2039 = **14/5** = The Temperance Card

This is the card ruling the sexually active person and can result in pregnancy, so it is quite possible Diana became pregnant during this time. Note also the crown concealed by the sun peeping over the horizon in the background of the card. The androgynous angel blends water and wine, or oil and water, pouring the two disparate liquids from one vessel to another, signifying the blending of seemingly inharmonious opposites. This card often symbolizes marriage between two people with different religions, different cultures or backgrounds, particularly those at variance with each other. The Number 39 though which the 14 is mediated, equates with the Knight of Cups, rules a relaxing romantic interlude and holidays to warm, summery places.

Matches for: Diana pregn ant third son die road side hop ital

Match: 1 of 2 Skip to Match

ELS (Equidistant Letter Spacing) Matrix

```
SOEVERHEGOETHTHESEWEREPURCHASEDFROMAMONGMENTOB
ETHEFIRSTFRUITSUNTOGODANDUNTOTHELAMBANDINTHEIR
MOUTHWASFOUNDNOLIETHEYAREWITHOUTBLEMISHANDISAW
ANOTHERANGELFLYINGINMIDHEAVENHAVINGETERNALGOOD
TIDINGSTOPROCLAIMUNTOTHEMTHATDWELLONTHEEARTHAN
DUNTOEVERYNATIONANDTRIBEANDTONGUEANDPEOPLEANDH
ESAITHWITHAGREATVOICEFEARGODANDGIVEHIMGLORYFOR
THEHOUROFHISJUDGMENTISCOMEANDWORSHIPHIMTHATMAD
ETHEHEAVENANDTHEEARTHANDSEAANDFOUNTAINSOFWATER
SANDANOTHERASECONDANGELFOLLOWEDSAYINGFALLENFAL
LENISBABYLONTHEGREATTHATHATHMADEALLTHENATIONST
ODRINKOFTHEWINEOFTHEWRATHOFHERFORNICATIONANDAN
OTHERANGELATHIRDFOLLOWEDTHEMSAYINGWITHAGREATVO
ICEIFANYMANWORSHIPPETHTHEBEASTANDHISIMAGEANDRE
CEIVETHAMARKONHISFOREHEADORUPONHISHANDHEALSOSH
ALLDRINKOFTHEWINEOFTHEWRATHOFGODWHICHISPREPARE
DUNMIXEDINTHECUPOFHISANGERANDHESHALLBETORMENTE
DWITHFIREANDBRIMSTONEINTHEPRESENCEOFTHEHOLYANG
ELSANDINTHEPRESENCEOFTHELAMBANDTHESMOKEOFTHEIR
TORMENTGOETHUPFOREVERANDEVERANDTHEYHAVENORESTD
AYANDNIGHTTHEYTHATWORSHIPTHEBEASTANDHISIMAGEAN
DWHOSORECEIVETHTHEMARKOFHISNAMEHEREISTHEPATIEN
```

Verses in Matrix

The Revelation to Saint John 14:4-14:11

Hopital = French for Hospital

Matches for: Diana Wales pregn ant die road side near hop ital

Match: 1 of 1 Skip to Match

ELS (Equidistant Letter Spacing) Matrix

```
RIESTSHALLCOMMANDTHATTHEYEMPTYTHEHOUSEBEFORETH
EPRIESTGOETHINTOSEETHEPLAGUETHATALLTHATISINTHE
HOUSEBENOTMADEUNCLEANANDAFTERWARDTHEPRIESTSHAL
LGOINTOSEETHEHOUSEANDHESHALLLOOKONTHEPLAGUEAND
BEHOLDIFTHEPLAGUEBEINTHEWALLSOFTHEHOUSEWITHHOL
LOWSTREAKSGREENISHORREDDISHANDTHEAPPEARANCETHE
REOFBELOWERTHANTHEWALLTHENTHEPRIESTSHALLGOOUTO
FTHEHOUSETOTHEDOOROFTHEHOUSEANDSHUTUPTHEHOUSES
EVENDAYSANDTHEPRIESTSHALLCOMEAGAINTHESEVENTHDA
YANDSHALLLOOKANDBEHOLDIFTHEPLAGUEBESPREADINTHE
WALLSOFTHEHOUSETHENTHEPRIESTSHALLCOMMANDTHATTH
EYTAKEOUTTHESTONESINWHICHTHEPLAGUEISANDCASTTHE
MINTOANUNCLEANPLACEWITHOUTTHECITYANDHESHALLCAU
SETHEHOUSETOBESCRAPEDWITHINROUNDABOUTANDTHEYSH
ALLPOUROUTTHEMORTARTHATTHEYSCRAPEOFFWITHOUTTHE
CITYINTOANUNCLEANPLACEANDTHEYSHALLTAKEOTHERSTO
NESANDPUTTHEMINTHEPLACEOFTHOSESTONESANDHESHALL
TAKEOTHERMORTARANDSHALLPLASTERTHEHOUSEANDIFTHE
PLAGUECOMEAGAINANDBREAKOUTINTHEHOUSEAFTERTHATH
EHATHTAKENOUTTHESTONESANDAFTERHEHATHSCRAPEDTHE
HOUSEANDAFTERITISPLASTEREDTHENTHEPRIESTSHALLCO
MEINANDLOOKANDBEHOLDIFTHEPLAGUEBESPREADINTHEHO
USEITISAFRETTINGLEPROSYINTHEHOUSEITISUNCLEANAN
```

Verses in Matrix

The Third Book of Moses, commonly called Leviticu 14:36-14:45

The verses above read as Quote: "And the priest shall command that they empty the house, before the priest goeth in to see the plague, that all that is in the house be not made unclean and afterward the priest shall go in to see the house and he shall look on the plague and, behold, if the plague be in the walls of the house with hollow streaks, greenish or reddish, and the appearance thereof be lower than the wall then the priest shall go out of the house to the door of the house, and shut up the house seven days.

And the priest shall come again the seventh day, and shall look and, behold, if the plague be spread in the walls of the house then the priest shall command that they take out the stones in which the plague is, and cast them into an unclean place without the city and he shall cause the house to be scraped within round about, and they shall pour out the mortar, that they scrape off, without the city into an unclean place and they shall take other stones, and put them in the place of those stones and he shall take other mortar, and shall plaster the house.

And if the plague come again, and break out in the house, after that he hath taken out the stones, and after he hath scraped the house, and after it is plastered then the priest shall come in and look and, behold, if the plague be spread in the house, it is a fretting leprosy in the house it is unclean.

And he shall break down the house, the stones of it, and the timber thereof, and all the mortar of the house and he shall carry them forth out of the city into an unclean place." Unquote.

Section 34. The Pitié-Salpêtrière Hospital where Diana died.

The Chapel of the Pitié-Salpêtrière hospital where Diana died. Quote: The Pitié-Salpêtrière Hospital is a teaching hospital located in Paris, France. It is known as the place where Diana, Princess of Wales, died in August 1997, following a car crash.

Salpêtrière was originally a gunpowder factory ("salpêtre" being a constituent of gunpowder), but was converted to a dumping ground for the poor of Paris. It served as a prison for prostitutes, and a holding place for the mentally disabled, the criminally insane, epileptics, and the poor; it was also notable for its population of rats.

In 1656, Louis XIV charged the architect Libéral Bruant to build a hospital on the location of the factory, founding the Hospice de la Salpêtrière. The building was expanded in 1684. By the eve of the Revolution, it had become the world's largest hospital, with a capacity of 10,000 patients plus 300 prisoners, largely prostitutes swept from the streets of Paris. From La Salpêtrière they were paired with convicts and forcibly expatriated to New France.

During the September massacres of 1792, the Salpêtrière was stormed on the night of 3rd/4th September by a mob from the impoverished working-class district of the Faubourg Saint-Marcel, with the avowed intention of releasing the detained street-girls; 134 of the prostitutes were released; twenty-five madwomen were less fortunate and were dragged, some still in their chains, into the streets and murdered. Madame Roland, a Girondin supporter of the Revolution in its first liberalising stages, recorded in her Memoirs that the Revolution "has been stained by villains and become hideous".

1857 lithograph by Arman Gautier, showing personifications of “dementia, megalomania, acute mania, melancholia, idiocy, hallucination, erotic mania and paralysis” in the gardens of the Hospice de la Salpêtrière.

In the first half of the 19th century, the first humanitarian reforms in the treatment of the violently insane were initiated here by Philippe Pinel, friend of the Encyclopédistes; his sculptural monument stands before the main entrance in Place Marie-Curie, Boulevard de L'Hôpital. Later, when Dr. Jean-Martin Charcot took over the department, the Salpêtrière became known as a psychiatric centre.

Charcot is often credited as the founder of modern neurology. Students came from all over Europe to listen to Charcot's lectures. Among them was a young Sigmund Freud. The Pitié-Salpêtrière is now a general teaching hospital with departments focusing on most major medical specialities.

In front of the main entrance to the Hospital, there is a large bronze monument to Philippe Pinel, who was chief physician of the Hospice from 1795 to his death in 1826. The Salpêtrière was, at the time, like a large village, with seven thousand elderly indigent and ailing women, an entrenched bureaucracy, a teeming market and huge infirmaries. Pinel created an inoculation clinic in his service at the Salpêtrière in 1799 and the first vaccination in Paris was given there in April 1800. End of Quote.

Reference:http://en.wikipedia.org/wiki/Piti%C3%A9-Salp%C3%AAtri%C3%A8re_Hospital

What were my feelings upon learning of the tragic history of the Salpêtrière? It was a place for the criminally insane, prostitutes, “fallen women”, the poor. The verses in the Bible, like the hospital itself in times past, point to a place of leprosy, plague, infested with rats. Diana lived her life in compassion for the suffering. She nursed people and babies with Aids gloveless, kissed them. Christ lived, worked and died among lepers and prostitutes, so there is no disgrace in dying in a place with such history. Far from it.

In all likelihood, if she had to choose her place of death, Diana would have no problem at all dying in the modern Salpêtrière teaching hospital, even knowing its sad past, because of her compassion and understanding of the poor and the suffering.

Yet there exists the possibility that there is another, more sinister side to her being taken to that particular hospital in her dying hours. If there was a plot to murder her, the taking of her to such a place may have been planned by those with twisted, sick minds. The Biblical verses above, (if the Bible Code is true), could be interpreted as pointing an incriminating finger at those who planned “retribution” against her, believing her to be acting immorally, insanely, even criminally, treasonably so.

Diana with young victims of landmines.

Prince Harry, his mother's son, in Africa.

William and Kate. Living and loving as Diana would have wanted. William playing polo 2007.

INQUEST REOPENS IN EDINBURGH'S COURT OF SESSIONS

Court appeal over Diana 'murder plot' by PETER ALLEN, Daily Mail Friday, August 10, 2012. http://www.dailymail.co.uk/news/article-204141/Court-appeal-Diana-murder-plot.html Comments (0) Author's note: there were no public comments recorded at 10.44am, on August 10, 2012, (date of publication), with Daily Mail stating that it "would be taking no more comments." In other words, the public has been disbarred from making any comments at all on this article. Now that the Queen's Diamond Jubilee celebrations have passed, and the London Olympics are well under way, we might get some press feedback on what is happening with the inquest into Princess Diana's and Dodi Fayed's deaths. Subjudice might be involved if anyone had been charged with a crime, but why all the secrecy about an inquest into the circumstances surrounding the death of Princess Diana (mother of heirs to the English throne), and her partner, Dodi Fayed?

Quote: Mohamed Al Fayed's sensational claims that Princess Diana and his son Dodi were murdered were put to a judge yesterday. The method mirrored an earlier MI6 plan drawn up to get rid of former Yugoslav president Slobodan Milosevic. And the motive was Establishment fears that Diana was planning to marry Dodi, Edinburgh's Court of Session (pictured above) heard. Richard Keen QC told Lord Drummond-Young that Diana and Fayed were being spied on by MI6, and that an official statement denying this was simply untrue. The US National Security Agency had confirmed the Princess was the subject of monitoring by American agents, possibly on behalf of MI6, he said. He was speaking on behalf of Fayed senior, the Harrods owner who is citing European human rights legislation to demand a reinvestigation of the crash that killed the lovers. He is presenting his case in Edinburgh because he describes his Scottish estate at Balnagown Castle, Ross-shire, as his "first" British home, and claims that gives him the right to Scottish jurisdiction. Mr. Keen described the police investigation of the Paris crash as "Clouseau-esque in its opaqueness".

Driver 'may have been informant'

Driver Henri Paul, who also died, was found to be well over the drink-drive limit and on antidepressants. But yesterday, the first day of what is expected to be a weeklong hearing, the court was told that Paul may have been an MI6 informant. Mr. Keen claimed Diana and Princes William and Harry were being "monitored" from around July 10, 1997, when they arrived at the Fayed estate in St Tropez in the South of France. By August, there was intense speculation that she was preparing to announce her engagement to Dodi.

Mr. Keen said that former MI6 agent Richard Tomlinson had confirmed Diana was being spied on. The retired spy had also claimed that MI6 paid a "long standing informant" at the Ritz Hotel in Paris, where Paul earned just £20,000 a year as assistant head of security. Referring to Paul's finances - he had more than £150,000 in the bank - Mr. Keen said: "Subsequent to the crash (it was found) he maintained 13 different bank accounts in Paris... which might suggest that he had some form of part-time job".

'Flash' in tunnel

Tomlinson also claimed reports of a "blinding flash" in the tunnel prior to the crash mirrored MI6 plans he had seen in 1992 for the attempted assassination of former Yugoslav President Slobodan Milosevic. Mr. Keen outlined other matters which, he said, led Al Fayed to the "reasonable belief that the life of his son Dodi may have been taken by force". The first, he said, was the "highly unusual route" taken by Paul as he drove from the Ritz to Dodi's apartment. The court was shown aerial photographs of Paris with two places highlighted where the Mercedes appeared to have been prevented from turning off where it might have been expected to. Mr. Keen said there were ten traffic cameras on the route between the Ritz and the Alma tunnel, none of which, Al Fayed had been told, were operational at the time of the accident.

A white Fiat Uno and a large dark motorcycle with two male riders had pursued the Mercedes into the tunnel and then overtaken it, he said. "At which point," Mr. Keen said, "two independent observers entering the tunnel from different directions reported a large, indeed enormous, radar-like flash of light in the tunnel." Mr. Keen said it was later found that traces of paint and rubber taken from the scene could be identified as coming from a Fiat Uno, which had prompted French investigator Judge Herve Stephan to call for an investigation. The fact that the judge later "countermanded those instructions" might "appear a little unusual", he said.

Diana's fears

The barrister said Diana herself had raised fears of a murder plot in a letter in October 1996 - the letter recently revealed by her former butler Paul Burrell. In it she wrote that "an accident" was being planned in her car. She mentioned "brake failure and serious head injury".

Mr. Keen said Dodi would have married Diana, and so become stepfather to Prince William. "If the Princess of Wales's own fears had one ounce of truth in October 1996, one is entitled to ask how much greater they may have been in August 1997 when it was generally anticipated a person openly denigrated by sections of the Establishment was about to become the stepfather to the future king."

The Lord Advocate, Colin Boyd, has ruled that the crash is outside Scottish jurisdiction, a verdict which prompted Al Fayed's challenge. Outside court Al Fayed said: "The most beautiful woman in the world was murdered with my son. I have been fighting for six years but I can see the light and justice can be done." End of Quote.

Section 35. The adoration of Diana, grown after her death, and how she has been elevated by many to the status of a saint and goddess, even thought to have endured a ritual sacrifice in the Alma Tunnel.

There have been numerous articles written on the supposed "sacrificial" nature of the alleged assassination of Diana - some fantastical - others a little more believable. Here is one of the more feasible articles, claiming that the Pont d'Alma was the site of ancient sacrifice to the Goddess Diana.

This one interested me, after I found meaning in the history of the hospital where Diana was taken after the accident. Some people have claimed that there were no less than five closer hospitals than the one to which she was taken; almost as if the ambulance went out of its way to deliver her to the Pitié Salpêtrière. If we conjoin the two theories, perhaps some of it makes sense.

Quote: The city of Paris was built by the Merovingians, a medieval dynasty which ruled France for numerous generations. Before converting to Christianity, the Merovingian religion was a mysterious brand of paganism. The Pont D'Alma Tunnel was a sacred site dedicated to the Moon Goddess Diana, where they used to practice ritual sacrifices. During those ceremonies, it was of an utmost importance that the sacrificed victim died inside the underground temple.

The assassination of Diana was a re-enactment of this ancient pagan tradition. Shortly after Diana was killed, Rayelan Allan (a researcher of esoteric history since the early 1970's, who was also married to Gunther Russbacher, a CIA/ONI operative) wrote an article called "Diana, Queen of Heaven". The article was picked up by numerous newspapers across the United States and Europe. However, no one fully understood the deeper meaning of the article. Therefore, Rayelan decided to expand it into a book.

The book states that in pre-Christian times, the Pont d'Alma area had been the site of a pagan temple of the goddess Diana and a direct gateway to heaven. Mindful of this safety net, the place was chosen by the Merovingian kings (AD500-751) to fight their duels, with the loser going directly to paradise. "Pont" means "bridge" and "Alma" means "soul" and for Merovingians, the site was a bridge across the "river of souls". So, Pont de L'Alma, the site of the accident which killed Princess Diana, means "Bridge of the Soul." End of Quote.

Reference: http://vigilantcitizen.com/vigilantreport/princess-dianas-death-and-memorial-the-occult-meaning/

Others found meaning in these Quatrains of Nostradamus, which they have interpreted thus:-

Quatrain 10.35
"Women's apparel in the Temple of Diana:
Going to be murdered by the unknown one from Marne".

This interpretation could score a point. Marne is not only associated with Champagne. It has a long history with the French Army, which has its training ground there. The Camp Militaire de Mailly straddles the border of Aube in the south while the Camp de Mourmelon occupies a large area north of Châlons-en-Champagne. Britain too has a military training ground in Marne, located on Salisbury Plain.

Reference:http://en.wikipedia.org/wiki/Marne

The "Temple of Diana" could be said to be Althorp, which houses a massive display of her "women's apparel".

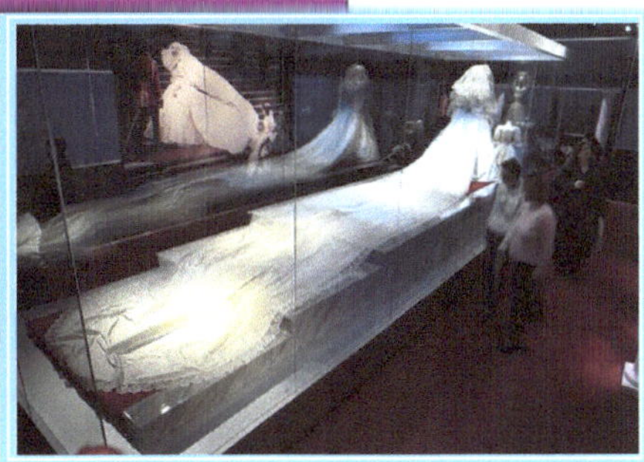

Century II. Quatrain 28.
"The last son of the man with the prophet's name
Will bring Diana to her day of rest
At a distance they wander in frenetic grief
Delivering a great people from subjection."

Dodi was the only son of Mohamed Al Fayed, and thus "the last son of the man with the prophet's name".

www.ingramcontent.com/pod-product-compliance
Lightning Source LLC
LaVergne TN
LVHW070952210926
848482LV00002B/195